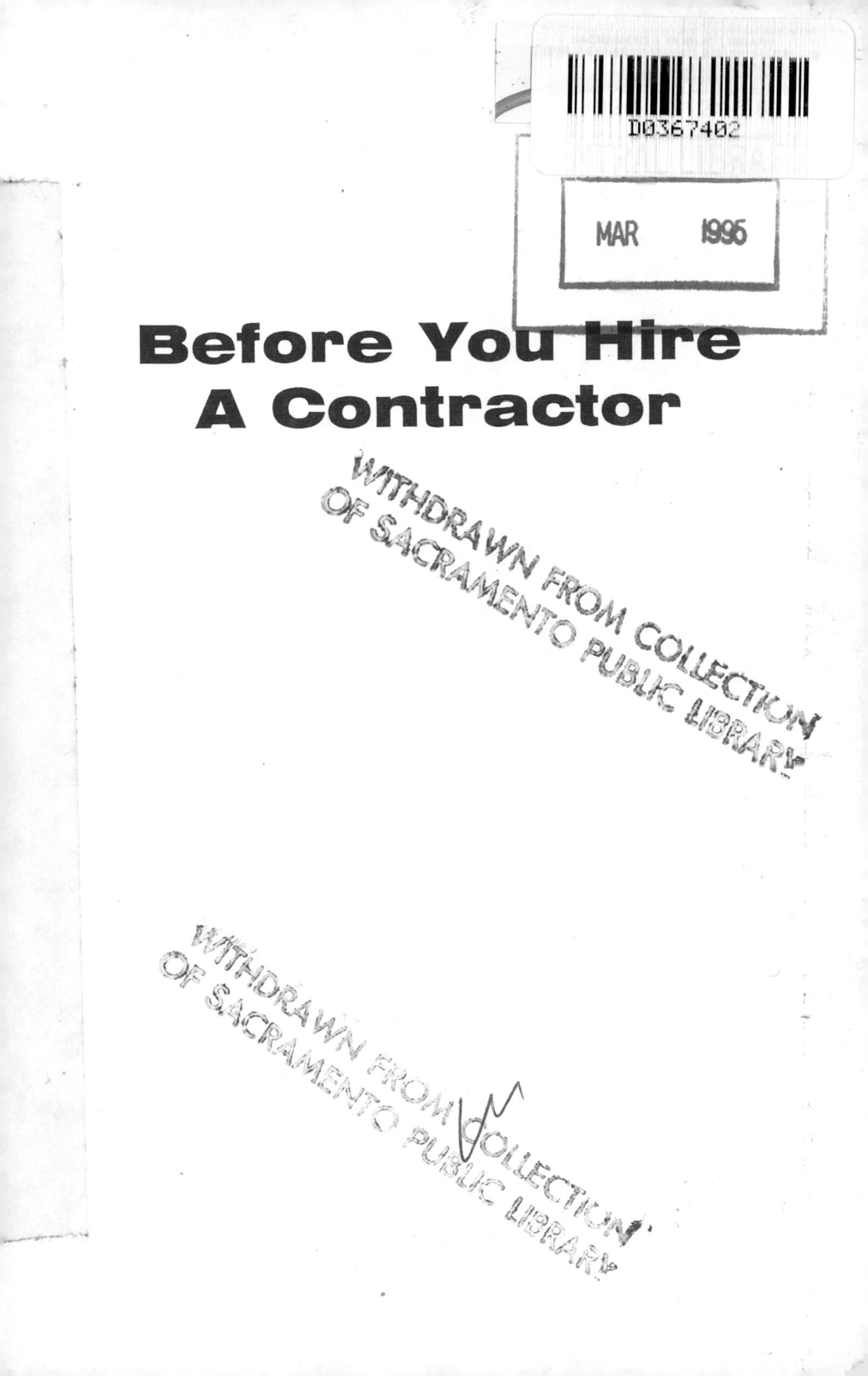

Before You Hire A Contractor

Before You Hire A Contractor

A Construction Guidebook for Consumers

by Steve Gonzalez, C.R.C.

Consumer Press
Fort Lauderdale, Florida

To all the readers who have benefitted from my project planning system.

Published by:
Consumer Press
13326 Southwest 28th Street, Suite 102
Ft. Lauderdale, FL 33330-1102 U.S.A.

Library of Congress Cataloging in Publication Data
Gonzalez, Steve, 1959-
Before you hire a contractor: a construction guidebook for consumers/ by Steve Gonzalez. — 1st ed.
p. cm.
Includes index.
1. House construction. 2. Contracting out. 3. Consumer education.
I. Title.
TH4812.G66 1994 94-37497
690'.837—dc20 CIP

ISBN 0-9628336-6-5 $12.95 Softcover

10 9 8 7 6 5 4 3 2
Printed in the United States of America

CONTENTS

Acknowledgements

Special thanks to all of the tradespeople and professionals who shared their knowledge, expertise and time to help create this book. They include:

Diana Hunter, Pam Price, Joseph Pappas, Diane Lentini, Chris Nicholson, Garry Spear, Laura Sotera, Regina Thomas, Pamela Jones, Colleen Tressler, Michael Haslet and Carole Collins.

Introduction

Getting involved. That's what it takes to get your project done the way you want it done, with lasting quality. I wrote this book to teach consumers *how* to get involved. The information provided will help you take charge of your construction project.

Having spent the past twenty years working in the construction industry, I have heard countless homeowners and contractors complain about each other. Part of the problem is that too many people put the same amount of trust in a contractor as they do their family doctor. They allow contractors to make choices for them and often have unrealistic expectations of how the project will turn out.

There is an alternative. *Do research and write specifications.* Get educated and apply your knowledge. From your research you will be able to write specifications of exactly what *you* want. You will no longer need to let a contractor make choices for you. Doing research also helps you understand the project's stages and lets you communicate better with a contractor.

Getting involved, doing research and writing specifications can all help you to achieve "the perfect project." This book will show you how.

Steve Gonzalez, C.R.C.

Chapter One

Project Planning: An Overview

Planning is an essential part of the building process. Unfortunately, it is often overlooked. Whether you want to remodel an existing floor plan, add a room or build a new home or office, project planning is important.

There are several steps involved in project planning. These steps often include hiring construction professionals. Your project planning may involve an architect, design-build firm, builder or contractor, or a combination of these.

Architects are design specialists who draw plans or blueprints. Some architects assist in selecting a contractor or builder. Many include project supervision in their fee.

Design-build firms perform every phase of a project, from the initial design to completion. They are often owned by architects who have a contractor's license. Many design-build firms have on-staff architects, engineers, builders and contractors.

Builders generally build new residential construction (townhouses, single family homes) or commercial construction (shopping centers, office buildings, restaurants).

Contractors specialize in a specific service, such as painting or roofing. They usually provide both labor and materials. **General and residential contractors** oversee entire projects, coordinating subcontractors, workers and supply deliveries. They may also act as the owner's agent, if authorized. Throughout this book, references made regarding contractors shall also apply to builders.

A System That Works

When it comes to planning, many people feel that a contractor or architect will "take care of everything." The fact is that getting involved plays a key part in acquiring "the perfect project."

The following steps are part of an effective project planning system that really works. These steps will afford you the opportunity to take an active part in planning your project. In addition, this system may help you to save both time and money. It can also greatly increase the potential for getting your project done properly and on schedule. Keep in mind that some steps may need to be done at different times during your project planning, depending on which professional services you use or other factors.

1. **Gather Information.** Write down a basic idea of the project. Collect magazine clippings, newspaper pictures, photos and samples. Start a file for notes, receipts, samples, lien releases, warranties and other information. Make drawings or sketches of what you have in mind. If you're not an artist, don't worry — having some idea of what you want is better than having no idea at all. Also make a list of the things you *don't* want.

Since materials make up the basis of a project, it's best to educate yourself about various products. Attend home shows and get product information on items and materials you'd like used for your project. Visit manufacturers' or distributors' warehouses and showrooms or write to them for illustrated information packets. Read construction magazines. Many of them are filled with advertising for new and innovative products. Some of these magazines have inquiry cards that you can mail to manufacturers to receive additional product information.

Many companies will be happy to send you product samples. Hold on to samples of all materials you decide to use for your project. They can be used for ordering as well as for comparison during installation.

Confirm proper installation procedures, how to care for the items or materials after installation and what type of warranty is given. Since improper methods of installation can void warranties, be sure to read them carefully *before* installation. Confirm that your contractor's method of installation meets the manufacturer's requirements. Remember to provide the contractor with a copy of the installation procedures for all items you supply.

Ask friends or acquaintances who have had similar work done about the materials used for their projects. It's especially helpful to ask about materials that get constant use, such as doorknobs, faucets and carpeting. Be specific — ask questions such as "Have your cabinets warped or delaminated? Have your roof shingles mildewed? Is your carpeting holding up well?" This should give you some idea of which products offer the best quality.

Keep in mind the fact that quality can mean different things to different people. Your idea of quality and your contractor's idea of quality may not be the same. Beware of clauses in contracts that state the contractor's right to substitute items "of equal quality" or "equivalent value." You may want to eliminate these clauses from your contract. Some owners add a clause requiring written notice and their signature before any changes are made. The clause should state that ample time (one week is common) will be given to choose a replacement. One note of advice here: avoid requesting materials and appliances that are not in stock, or are no longer produced by the manufacturer. Be sure your contractor knows where to get the items you request. Another note of advice: if using an architect, be sure your contract with the architect does not allow him to evaluate changes and make substitutions on your behalf without your written consent.

Not only can product quality differ from company to company, it can also vary within each brand name. Most brand name product lines have good, better and best product selections. Check them out!

These ideas will help you to select and specify the exact materials you want. You can also go to your local library and bookstores and ask for books and videos on your type of project. *Hometime®* offers several excellent lowcost videos. You can obtain a catalog of their videos by writing to: Hometime, Order Department BHC, 4275 Norex Drive, Chaska, MN 55318.

2. **Obtain a Survey.** A survey is a drawing of your property showing where all structures, trees, wells and utilities are located. In most cases, a survey for a home or parcel of land is presented to the buyer upon purchase.

After locating your survey, check the date it was drawn. Building and regulatory departments usually require a survey to be less than three years old. These may include the health department, environmental control board, zoning department and building department. If your survey is three years old or more, and you need to obtain a new one, contacting the original surveyor may save you both time and money.

By obtaining copies of the survey yourself, you will: a) know the status of your land according to the survey; b) have the document in hand to present to all parties who will require one; c) be likely to save both time and money by acquiring it yourself; d) take an active part in your planning.

Be sure to obtain several copies of your survey *that have the surveyor's seal* to give to all parties that will require one. Your building department and mortgage company may require additional surveys throughout the course of your project. They may also require a final survey or elevation certificate upon completion of your project. Ask your surveyor for a price on a "package deal" for all required surveys and certificates.

3. **Visit Your Homeowner's Association.** Ask for a detailed list and breakdown of all restrictions and requirements that affect your home (including any upcoming changes and addendums). Carefully read them and question any misunderstood sections. If necessary, ask an attorney for assistance.

Homeowners' associations may restrict home size, paint and stucco color, window types, number of windows or doors, landscaping and several other aspects of your home (along with the size of your pets and the number of kids you can have). They may also restrict where a new home can sit on a piece of land, which direction a home faces, or whether it can be a two-story residence. Many associations restrict home offices. *Even if your city or town allows a home office, your homeowner's association might not.* Be sure to check.

Some homeowner's associations have architectural control boards that regulate these restrictions. If you will be required to present your plans for review and approval, confirm how long the process will take. Keep in mind that there may be a fee for the homeowner's association to review your plans. Also be aware that various homeowner's associations may require you to provide an amount of money to be held until your project is completed. The purpose of this may be to ensure that you will comply with all of the association's regulations. Knowing about additional money requirements in advance will help you to avoid unnecessary surprises.

4. **Visit Applicable Building & Zoning Departments.** Bring your survey and a rough drawing of the project, including the approximate size and dimensions. Ask about easements, dedications, setbacks, rights-of-way, utilities, future roadways and any other restrictions that may prevent you from proceeding with your project. Find out the requirements of an architectural control board, if your area has one. Confirm whether the building, zoning or other applicable departments will charge a fee to review your plans once submitted. Do the same for architectural control boards.

   While you're there, talk to the chief building inspector or plans examiner. Inquire about new or upcoming building requirements.

   If zoning restrictions will not allow you to build or remodel the project you have in mind, you may have two options: 1) if the project is restricted due to size, reduce the dimensions or consider a two-story structure (if permissible); or 2) ask for a variance, special use permit or waiver.

Find out the legal requirements that must be met in order to be granted a variance or special use permit. If necessary, ask an attorney to represent you before the zoning board. *Be sure to confirm that the changes will be allowable according to your homeowner's association restrictions before applying for a variance or special use permit.*

You may need to present a written appeal to the zoning board when asking for a variance. It is usually helpful to obtain several written approvals and signatures of immediate neighbors not opposed to your construction plans. Present these along with your written appeal. Reasonable variance requests are often granted. This is common in cases where the property owner has complied with all necessary requirements and there is no opposition among area homeowners.

The length of time necessary to receive a variance can vary greatly. Factors to consider include the number of other variance requests being processed, the number of persons or boards involved in the approval procedure and other factors. It is important to confirm how long it will take to receive an answer to your variance request. For example, if it will take eight weeks to provide an answer, you may want to place the project on "stand by." If you are told your variance "will probably be approved," it is not advisable to continue with any type of project planning that requires money. In most cases it is best to collect information while waiting for a *written* variance approval. You may be required to attend a city or town council meeting for an approval.

5. **Select a Draftsman for Preliminary Plans.** After confirming the requirements of your local building department and homeowner's association, the next step is to obtain *preliminary* plans. Preliminary plans are the first set of blueprints drawn for a project. After all changes and additions have been made, final plans are produced.

   Plans can become the basis for which all members of your project "team" coordinate ideas, prices and suggestions. Most importantly, plans help you prepare written specifications of work to be done and materials to be supplied. Preliminary plans are often part of a "package deal" which includes other services.

   Architectural contracts may include preliminary and final plans, revisions, product selections, landscape and design services, engineering, drawings for future expansion and project supervision. Some contracts include only a portion of these services.

   Many people purchase plans through magazines or blueprint outlets. These plans often need to be altered before being used for a project. Altering can sometimes cost more than drawing plans from scratch. Get a price, *in writing*, for any alterations needed or required before having the plans changed.

Many blueprint companies charge extra fees for diagrams of recommended wiring, installation procedures and project material lists. Since every project is unique, there may be no guarantee that lists of required materials will be complete. Whatever the case, it's best to deal with blueprint companies who offer a money-back guarantee (P.S. — read the fine print *before* you order).

If you will be considering an architect to design a new home, be sure the person has experience in *residential* design. *There are more commercial architects than residential architects*. Most residential architects subscribe to architectural and building magazines that profile styles of homes from all over the world. This makes many residential architects familiar with a large variety of floor plans.

Coordinating your needs and desires with an architect's experience in custom design can help you obtain a floorplan that is both stylish and practical. To best do this, an architect will need you to answer certain questions about your lifestyle. Listed below are just a few:

1) What is your budget?
2) How many bedrooms do you need?
3) How many people are in your family?
   What are their ages?
4) How long do you intend to live in the home?
5) Do you have pets? Will they live indoors,
   outdoors or both?

6) Will a maid or in-law be living with you?
7) How many cars and other vehicles will you have?
8) Do you want a garage, attic or basement?
9) Will there be a fireplace?
10) Do you want a pool?
11) Do you want a patio? Will it be enclosed?
12) Is there a need for handicap design?
13) What type of laundry room would you like?
14) Where would you like the kitchen and bathrooms to be located? What size would you like them to be?
15) What are your storage requirements?

Be as specific as possible when answering questions and discussing each room's design. The more specific you are, the better an architect can interpret your needs.

When interviewing architects, ask if they are current on the cost of all types of new materials and how they should be installed. Also ask whcther they have been trained in engineering. Most building departments will accept structural plans drawn by either an architect or engineer. Others require a structural engineer to review the structural parts of a project.

Plans drawn by designers or contractors are usually required to be reviewed and approved by a certified engineer. Part of an engineer's job is to review plans for code compliance. Engineers often recommend changes to plans that will make a structure sturdier or safer. Of course, if an engineer recommends costly changes that you feel may be unnecessary, consider getting a second opinion.

Whether you select an architect, engineer or contractor for your preliminary plans, a qualified professional should be hired. Carefully check the credentials of each individual you interview. Ask for copies and confirm the validity of all licenses, insurances, registrations and bonds. Also ask if they carry errors and omissions insurance for coverage in the event of design errors. A major design error could cost more in repairs than the entire project. Remember to confirm *in writing* from each potential draftsman the length of time it will take to complete the plans.

Obtain the names and telephone numbers of several references who have had plans drawn by each draftsman. Ask for referrals whose projects were similar to yours. Call the references and ask if there were any problems in structure or design as a result of the drawings.

6. **Write Specifications.** Specifications are a highly important part of any contract agreement. Obtain as many detailed specifications as you can from the person drawing your plans. For a complete description of what specifications are and how to obtain them, *see Chapter Three*.

7. **Interview Contractors.** If you decide to use a design-build firm, you will probably choose your contractor as part of the design-build "package." This step will then need to be done before having your preliminary plans drawn. Also, if you have chosen a contractor to prepare your preliminary drawings, you may decide to use the same person for your final drawings and construction. If this is the case, you should have already interviewed the contractor, checked his references and requested copies of all license and insurance documents (*see Chapter Four*). However, if you have selected an architect and have preliminary drawings, you may now be in search of a contractor.

Whether you interview contractors referred by family, friends, business colleagues or an architect, be sure to obtain a minimum of three bids. Personally discuss your project **in person** with each contractor, not a "sales representative," "job planning specialist," or "agent." You will be spending a substantial amount of money with whoever you choose, so demand to have them personally interview with you. This also helps eliminate misunderstandings and helps you to get a general idea of how well you would be able to work together. **It is also the perfect time to present your specifications list and general conditions.** Ask for a copy of each contractor's written warranty for materials and workmanship.

When you interview each contractor, get as much information as possible. Ask how much time is needed to get back to you with a bid. When you receive each bid, tell the contractor that you will get back to him in about one week. This should allow you enough time to thoroughly screen his credentials and references, as well as giving you time to compare each bid submitted.

8. **Call a "Team Meeting."** Once you have chosen a contractor to work with, call and set up a meeting between yourself, the contractor, the architect and any other professionals who will be doing work on your project. These professionals may include interior or landscape designers and pool builders. The purpose of this meeting is to coordinate and confirm work schedules, hear comments and suggestions, voice expectations and tie up any other "loose ends."

9. **Obtain Completed Plans.** After hearing all of the comments and suggestions during the team meeting, you may find one or several items that you would like added, deleted or changed on your plans. This is why preliminary plans are often a good idea. It is generally much easier and more cost efficient to make changes now, rather than when the plans are completed. Several copies of the completed plans should be made (one for each department doing a review). Be sure to obtain one set of plans for your personal use.

10. **Negotiate the Contract.** There are usually two occasions during a project when you need to negoti-

ate a contract. The first time is at the design stage, when you hire a draftsman to draw your plans. The second time is when you hire a contractor to begin construction.

Your list of specifications should be included in each contract **before you sign**. Include commencement and completion dates, the total contract price, payment terms and general conditions. (*See Chapter Five*).

Consider having an attorney review your contracts. A real estate attorney with construction exerience who specializes in construction law (and doesn't have a personal or business relationship with your contractor) may be the best source. Your local bar association, courthouse or library may have a lawyer referral service.

NOTES:

Chapter Two

Scams, Problems & Nightmares: Avoiding the Maze

Almost everyone has heard a building project horror story. The staffs of many consumer protection agencies, Better Business Bureaus and attorneys' offices hear hundreds of construction complaints every year. Here are some examples of the problems encountered and ideas that may have helped prevent them:

The Situation

A homeowner hired an interior designer. She paid the designer a fee that included not only design services, but also the purchase of furniture, drapes and ornamentals. When the project was done, the homeowner realized her $5000 couch was defective. The interior designer, who had left town, never provided the homeowner with a receipt from the actual supplier of the couch.

How It Could Have Been Prevented

In her contract with the interior designer, the homeowner should have requested the make, model number, manufacturer and distributor of all furnishings. An original written receipt and warranty should have been required prior to payment for services.

The Situation

A newly married couple hired a builder to remodel a house they had just purchased. The builder provided them with a copy of his business and occupational licenses as well as his insurance documents. The couple researched the validity of his documents and everything checked out. Halfway through the project the wife found a wallet among debris. She opened it to find out who it belonged to and was shocked to see her builder's picture on a driver's license — with a totally different name.

How It Could Have Been Prevented

Several things could have been happening here. The builder might have been using someone else's occupational license and insurance (with or without their knowledge). Or the builder could have been carrying false identification. Despite which one, if legal problems arose the question would be "Who do I take legal action against?" This couple should have asked to see the builder's driver's license and requested a copy of it before they signed the contract.

The Situation

A couple hired an architect to draw blueprints for a family room addition. They interviewed a well-known, reputable builder and presented their blueprints. A few days later they signed a contract and issued a 5% deposit.

The builder spent days personally photographing and measuring the property for accuracy. He also acquired information and filled out forms for the required building permits (the home was located in a highly regulated area, part of which was a historical site). The builder then attended an architectural control board meeting. After presenting the information to the proper authorities, he was informed that the blueprints were not drawn in accordance with zoning restrictions.

The builder approached the homeowners and suggested having new blueprints drawn to conform with the zoning restrictions. He also suggested they consider seeking a variance. Instead, the homeowners decided to cancel the contract between themselves and the builder. The builder agreed, and requested cancellation in writing as well as payment for services rendered. The builder had already spent over three weeks organizing the project.

Upon presentation of the final bill, the homeowners requested a full refund of their deposit. They felt that since their project was not buildable, they should not have to pay any money to the builder. Neither the homeowner nor the architect checked with building and zoning prior to the blueprints being drawn.

How It Could Have Been Prevented

The homeowners should have visited their local building and zoning departments *before hiring an architect*. They also should have confirmed the requirements of the architectural control board. The homeowners should not have relied on the architect (who obviously didn't check) or the contractor (who *did* check and ended up losing the job).

The Situation

A tornado devastated a community. One homeowner hired a builder and his crew to rebuild her home. They were neat, timely and organized and produced a fine finished product.

The homeowner's neighbor, having difficulty in finding a builder, asked her to refer the builder. The builder agreed to rebuild the neighbor's home at a price lower than two other bids she had received. The contract was all-inclusive and the builder requested a fifty-thousand-dollar deposit. The homeowner paid it (good grief!). One week passed and the project ceased. The crew didn't return. When the owner contacted the builder, he offered his apologies and told her that his partner took the money and left town. Meanwhile, the woman did not have enough money to hire another builder. Her home was now more damaged and exposed to the elements than it was after the storm.

How It Could Have Been Prevented

The homeowner never should have made a deposit payment of such a substantial amount of money. Five to ten percent **upon commencement** is usually sufficient. A draw schedule, showing payments to be made as stages of work were completed, should have been included in the contract (*see Chapter Five*). Consulting with an attorney may have been helpful.

The Situation

A couple hired a developer to build their new home. The wife, excited about her new kitchen, carefully selected her new appliances. When the wife visited the project one day to see the completed kitchen, none of the appliances she requested had been installed. Instead, there were other appliances (in a different color). Upon approaching the builder, she was told that the contract specifically stipulated he could use "appliances of equal value." The builder stated that his regular supplier did not have them in stock and he did not want to delay the project.

How It Could Have Been Prevented

The couple's contract should have included a change order clause stating that the contractor must obtain written authorization prior to making changes. That way the owners could have checked if other suppliers had the appliances they wanted.

The Situation

A builder was hired to construct a $200,000 home. The construction began and within three months the home was completed. Within one week of moving in, the owners began receiving notices of liens and bills from suppliers and workers. The total of these bills was an additional $204,000. The owners visited the builder's office, only to find a vacant shop at the end of a plaza. Local tenants told them the "office" had been unoccupied for over a year.

How It Could Have Been Prevented

The owners should have received partial releases of lien as the project progressed. A final release of lien should have been obtained from all suppliers *and* the builder before making final payment (*see Chapter Six*). Visiting the builder's office before hiring him may have helped these owners avoid this scam. Confirming the builder's home address might have been helpful as well.

The Situation

A homeowner hired a contractor for a $20,000 kitchen remodeling project. The contractor requested a 20% deposit upon signing the contract. Six weeks later, the permit was approved and issued. The contractor approached the homeowner and asked for additional money before he would start the project.

How It Could Have Been Prevented

A deposit should have been given on the date the contractor presented the homeowner with a valid permit and commenced work. A financially secure contractor should be able to obtain permits and order supplies through supply accounts without receiving a large deposit.

The Situation

A homeowner hired a handyman to remove a wall from her dining area. Her dining room was soon filled — with the floor and contents of the room above it.

How It Could Have Been Prevented

The homeowner should have hired someone qualified, who could have recognized a load-bearing wall.

The Situation

A couple bought a parcel of land intending to have a new home built for their growing family. They signed contracts with a builder and a pool company. The pool company commenced work before the builder. The pool was constructed one foot from the back of the house.

How It Could Have Been Prevented

Procedure was the problem here. The pool should have been put in *after* the home construction was completed. This was a major planning blunder. The builder and the pool contractor should have coordinated the schedule with the owner and architect during the "team meeting."

The Situation

A couple decided to replace their old wooden kitchen cabinets with new mica cabinets. They called three reputable contractors with good references and obtained bids. One bid came in at $7,000, the second at $8,100 and the final bid at $9,500. Since all had good references, they chose the lowest bidder. While the cabinets were being delivered, the wife noticed that the insides were melamine, not mica. The contractor agreed to replace the insides with mica as well — for an additional $3,800. The project cost was now $10,800 — $1,300 more than the highest bidder. The highest original bid of $9,500 had included mica insides.

How It Could Have Been Prevented

Specifications! This couple may have assumed that all mica cabinets have mica insides, or they may not have considered it at all. Consider everything, assume nothing and always "get it in writing."

The Situation

An elderly man hired a builder for a small room addition. The builder told the man that getting his own permit would speed up the project and save him money. He convinced the man to obtain the permit himself. The builder also requested a 40% deposit on the $10,000 project. Then he took the man's deposit and left town.

How It Could Have Been Prevented

This man should not have pulled his own permit. Homeowners should *never* pull permits unless they are actually doing the work themselves (pulling a permit makes you liable for the project). He should have also checked the builders references and avoided giving more than a 5-10% deposit.

The Situation

A man hired a builder from a reputable, prestigious building firm to remodel his home. The builder had the homeowner make all payments payable to him personally, rather than the company he worked for. Two months later, the project was finished. Several liens were placed on the home. When the irate homeowner called the building firm, he was told that the builder had been fired over two months ago.

How It Could Have Been Prevented

The homeowner should have called the building firm to confirm the builder's employment. He should not have written checks to the builder personally. Also, the homeowner should have requested releases of liens.

The Situation

A homeowner hired a contractor to remodel his den, family room and dining room. The homeowner promised his brother six expensive ceiling fans that would be replaced during remodeling. Meanwhile, the contractor had bid another job that would include installation of those same six fans. The builder's contract stated that he may take possession of all replaced and surplus fixtures.

How It Could Have Been Prevented

The homeowner should have read the contract carefully and added a clause stating that all existing fixtures prior to construction remain the property of the homeowner.

The Situation

A couple hired a builder to build a nursery for their new baby. Upon completion of the job, the builder presented the couple with a signed permit card and requested his final payment. All appeared to be in working order, so the couple paid the builder. Two weeks later the couple was questioned by municipal authorities and issued a fine. The final signature on the permit card had been forged.

How It Could Have Been Prevented

A phone call or visit to the building department to confirm that the final signature was valid would have been in order here.

The Situation

An elderly woman was approached at her home by several workers in a white van. The workers presented a business card and offered her a "special" on refinishing her driveway. The special was for $200. The woman agreed, and the workers got underway. Two hours later, the supervisor of the group knocked on the door and presented the woman with an invoice for $850. The woman asked for an explanation, and was told that it took eight barrels more to complete her driveway than was included in the "special." The supervisor told the woman that she must pay now, in cash, or a lien would be placed against her home. The woman drove to the bank to obtain the money, and while there, the "workers" robbed her home. When she arrived with the money, they requested payment and drove away. Within a week, the driveway's cracks and crevices resurfaced.

How It Could Have Been Prevented

It's never a wise idea to deal with door to door solicitors when it comes to home improvements (and many other things for that matter). This woman should have asked for a business card and called the company for verification. If she got that far, the next step should have been to ask for a written estimate **with specifications** of the exact materials to be used and a firm price. Then she should have obtained at least two other estimates from qualified professionals.

The Situation

A couple hired a contractor to build a bedroom addition on their home. They did not check his references. The contractor rarely showed up, and when he did, drank alcoholic beverages while he worked. The contractor was sent to prison several times during the first few months. The project was only partially completed when the homeowners were informed that the contractor was being deported. Meanwhile, the project's permits expired.

How It Could Have Been Prevented

This couple should have thoroughly checked the contractor's references. A clause in their contract stating that no alcoholic beverages were to be consumed on the project may have been helpful. Another clause, requiring the contractor to be at the job site for a set number of days per week, may have made him show up more frequently.

The Situation

A man hired a contractor for a room addition. After the project commenced, the man came home from work one afternoon to find truckloads of construction debris in his back yard. He called the contractor and was told that the debris was from the contractor's other work sites. The contractor said that it would be removed the next time he went to the dump. The homeowner received many calls and letters from angry neighbors. He also had to put up with rodent infestation and fines from a homeowner's association.

How It Could Have Been Prevented

This homeowner may have benefitted from a clause in his contract that stated either: a) the job site would be cleaned and all debris removed on a daily basis, or b) debris would be stored in an approved dumpster and emptied regularly during the course of the project.

The Situation

A couple hired a builder to construct a single family residence in a rapidly growing area. They signed a contract for the construction contingent upon their qualifying for a specific lot. The couple was informed that they qualified, but were also slapped with $6,000 in impact fees.

How It Could Have Been Prevented

This couple should have inquired about impact fees before signing anything. Rapidly growing areas are a red flag for impact fees. These fees are used to help build new schools, parks, hospitals, and fire stations, among other facilities. Occasionally, impact fees are not issued until the permit is requested. When considering land, provide the legal description of the property to the local building and zoning departments and ask about impact fees. If you are told there are no impact fees, *get a copy of the confirmation in writing.*

The Situation

A man purchased blueprints from an ad in a magazine for $450. He then hired an architect to make changes to them. Two weeks later, the architect called the man and told him to pick up his blueprints. He was presented with a bill for three times the cost of the blueprints (that were intended to save him money).

How It Could Have Been Prevented

The man should have confirmed *in writing* a set price for the changes to the plans. It may also have afforded the man an opportunity to send the blueprints back to the outlet, if they carried a money-back guarantee and the company's policies allowed it.

The Situation

A couple hired a designer to draw plans for their new home without checking references. When the plans were completed, the couple paid the designer the fee they had agreed to in writing. The couple then hired a builder, who informed them that the plans were both incomplete and incorrect. The builder recommended that the couple find a good architect.

How It Could Have Been Prevented

The couple should have asked for references and checked them. They should have also asked about the designer's training and experience. In this particular case, the "designer" had simply taken a drafting course and had fancy business cards printed. Stating in their contract that a portion of the payment would be held back until the plans had been approved may have helped.

The Situation

A woman hired a contractor for a bathroom remodeling project. More than halfway through the project, the woman fired the contractor (with good cause). She hired another contractor to complete the project. Two months later, problems arose due to work from the first contractor. The woman called the second contractor, who informed her that his warranty did not cover the first contractor's work.

How It Could Have Been Prevented

This woman should have required the first contractor to have a performance and payment bond. If he had, the bonding company may have paid to have the work completed and warrantied it (*see Chapter Seven)*. Confirming the warranty of the second contractor would have eliminated the element of surprise.

The Situation

A couple bought a piece of land to have a home built. They hired a contractor, who informed them that excavation of the land would cost a substantial amount of money. The land was mainly solid limestone.

How It Could Have Been Prevented

A soil test should have been performed by a registered engineer before the land was purchased. The test, which costs an average of $250, shows what type of ground material lies beneath the surface. The test is commonly done using hollow core rods which are driven 10 to 20 feet into the ground. This is done by a hydraulic machine that looks similar to a large steel mallet. Samples are usually taken from all four corners of the proposed foundation. The rods are then taken to a laboratory and their contents are analyzed. The engineer then provides the homeowner with sealed copies of the soil statement.

The Situation

A couple hired a contractor to build a house. The contractor hired an excavator to bulldoze land fill that was scheduled to be delivered to the site. The excavator arrived at the job site with his bulldozer. He sat at the site for eight hours waiting for the fill to arrive. He did this for three days. The land fill was never delivered. The homeowners were charged for the three days the bulldozer operator waited at the site. The homeowners refused to pay the bill and a lien was placed against their property.

How It Could Have Been Prevented

In most cases, a contractor is responsible for the actions of his subcontractors. The contractor should have been on the project, or in touch with the subcontractors to find out the status of the project. The couple should have contacted the contractor and advised him about the situation. A clause in the contract stating that the contractor would be on site a set number of days per week may have helped.

The Situation

A couple hired a contractor for extensive home remodeling. During the course of the project, one of the contractor's employees made over $150 worth of phone calls to a foreign country.

How It Could Have Been Prevented

The couple should have added a stipulation in the contract stating that the phone was not to be used (*see Amenities Clause, Chapter Five)*. The stipulation should have included that any illegitimate calls charged to the owner's phone bill would become the sole responsibility of the contractor.

In Summary

In all these cases, one thing is clear: consumer education beforehand could have helped prevent problems. Doing research, adding clauses to your contract and staying involved with your project as it progresses can make a world of difference.

Proper inspections are also important. They can play a big part in ensuring that the job is done right. Most building inspections done by government officials are only to enforce code requirements. Unfortunately, many of these requirements only require the contractor to meet minimum building standards. Sloppy workmanship or defective materials may slip by without being corrected. Some towns, cities and counties have local requirements in addition to the standard building codes. Ask about local requirements and try to be on the job site when inspections are done.

Some building inspectors have limited training. Ask if the inspectors who will be inspecting your project are certified, and if so, by whom. Several types of inspectors are likely to inspect a project from beginning to end. These may include a plumbing inspector, electrical inspector, and structural/mechanical inspector, among others.

Overloaded building departments can cause problems such as high employee turnover, overworked employees and long waits for inspections. This may cause inspections that are not thorough enough, or the signing of inspection cards without an actual inspection. These can be potentially dangerous situations. Watch for signs of this type of activity.

Some people hire an independent inspector or construction consultant to perform inspections on their project. This can be especially helpful if you are building a new home or having renovations done while you're not in town.

An "independent" or "private" inspector usually has no affiliation with local building and zoning departments. A real estate or construction attorney may be a good source for finding someone reputable.

Before hiring an independent inspector, be sure to confirm that your homeowner's association and building departments will accept his inspections. If they won't, you may still want to hire an inspector to confirm that the work is being done correctly.

When selecting an inspector, be sure to screen him in the same way as the architect, contractor or other parties involved with your project. Remember to get references and check them. Ask for detailed documentation of his training and expertise. Find out if the inspector has ever made inspections in your area, and if he is familiar with *current* local requirements and codes.

Confirm the cost per inspection, the number of inspections that will be made, and the type of written reports you will receive. Require the independent inspector to inform you of inspection dates at least 2-3 days in advance (in case you'd like to be there). Also require photographs to be taken as each stage of the project is completed.

Confirm that your independent inspector may visit the project and has full access at any time. Check that the contractor's liability insurance does not restrict you or others working for you from coming on the job site. If it does, ask for an exception letter or waiver.

No matter who inspects your project, obtain *detailed* written documentation of approved or failed inspections *at the time of each inspection.*

NOTES:

Chapter Three

Specifications: The Key To Getting Exactly What You Want

Specifications are a vital part of any building or remodeling contract, no matter how small. Owner specifications *should accompany and become a part of every contract* before the contract is signed. The specifications should include all supplies to be used, all services to be provided and all terms and conditions to be abided by. In cases where the specifications list is added as an addendum to the contract, be sure there is a place for the owner and contractor to sign and date each additional page as acceptance of the terms.

Careful consideration and preparation of your project specifications is important. It can lessen or eliminate the possibility of your project going over budget due to the cost of extras and change orders.

Specifications are easily obtained by writing down all characteristics of materials and supplies for your project. Specifications should include the exact brand, make, model number, style, type, color, shade, size, weight, thickness and any other description. It is especially important to include easily overlooked items in the specifications. Some examples: 1) the brand of primer that you want on the walls before they are painted; 2) the brand, type and thickness of carpet padding for your carpets; 3) whether the window sills in each room will be wood, marble or another material. The specifications should also state that all materials used will be new, unless otherwise agreed to in writing.

An effective way to start your specifications list is to visualize each room on the plans. Write down all of the finishes and fixtures you would like on the walls, ceilings and floors. For example, if you are visualizing the walls, include the plaster, primer, paint or wallpaper, mouldings, windows and window sills. Don't forget paneling, niches, shelving or any other items you'd like. *Make sure you include the specifications in your contract.*

When building a new home, review your plans and visit homes similar to yours to get an idea of what to include. When choosing from model homes, include specifications for every item you see. Models that are all-inclusive may include a washer, dryer, dishwasher, self-cleaning oven and wood flooring. However, unless you specify the exact model numbers, colors and brands of what you see, you may receive ones of lesser quality, or a different style or color. You may also find that what is on display is available — at an additional cost.

Keep in mind that most model home builders will not accept a contract from a consumer. Many model home builders are hesitant to make changes to their own contracts. Some downright refuse. Also, they may or may not allow change orders after you sign a contract. *Be sure to check before you sign.*

Here is another instance where specifications are extremely important: Two reputable contractors bid on a remodeling project. Both include in their estimates Brand "A" casement windows. One contractor regularly uses Brand A's *top of the line* casement windows, while the other contractor uses Brand A's *standard quality* casement windows. This is a good example of why you need *exact* specifications.

Obtaining samples to compare during installation helps assure that you will get what you are paying for. You can obtain samples of flooring, tile, paint, wood, carpet, carpet padding and numerous other building materials by calling or writing to the manufacturers. You can also visit their showrooms. Home centers and home shows also offer an array of product samples. Reading building trade magazines will provide you with names and addresses of manufacturers to write to for product information.

The following is a list of many of the specifications used for materials:

A/C Systems Brand, make, model number, energy efficiency rating, size of heat strip.

Appliances Brand, make, model number, color.

Asphalt Thickness, brand and type of sealer to be used (if sealed).

Awnings Brand, type, model number, color, material (steel, aluminum, wood or other).

Backsplashes Brand, model, color, material (tile, natural stone, mica, metal or other).

Banisters Material (wood, metal, plastic, glass or other). If wood, be sure to include finish type.

Baseboards Material (plastic, vinyl or wood). If wood, what type, thickness? Will the wood be clear (one length) or finger joint (random length pieces, joined together)? Will it be painted, stained or left natural?

Bath Fixtures Brand, make, model number, color, material (porcelain, ceramic, natural stone, wood, glass, plastic or other).

Bathtubs Brand, make, model number, size, color, material (cast iron, steel, plastic or other), with or without jets. If motorized (i.e.: hot tubs), specify horsepower.

Blinds Make, model number, size, color, material (plastic, metal, cloth or other), trim type if included.

Cabinets Material (wood, mica or other material). If wood, specify type and finish. If mica, specify brand and model number. Include thickness of material, number of drawers, inside material, type of glass (if used), hardware, frames, trim, lighting and any special features.

Carpet	Brand, color, order number, density, pile, weight, heat treatment or other.
Carpet Padding	Weight, thickness, brand, material type (foam, nylon, rubber or other).
Ceilings	Material (plaster, wood, joint compound or other), texture, finish (popcorn, diamond dust or other, sprayed or plastered), type (tray, drop, vaulted, acoustic or other).
Ceiling Fans	Make, model number, color, light kit included (if applicable), reversible or not, number of blades.
Chimneys	Type of flue material (inside of chimney, usually clay or brick). Type of lid and damper, color.
Concrete	Thickness, type of sealer if sealed, color if stained, design if stamped. Include pounds per square inch (3000 psi is common for driveways, 2500 psi is common for slabs).
Countertops	Make, model number, color, thickness, material (solid surface, natural stone, mica, wood, ceramic tile or other).

Door Knobs

Brand, make, style, model number, handle type (round, square, lever or other), finish (polished chrome, brass, antique brass, brushed chrome, nickel, solid brass, plastic or other). Also specify if the locks are to be keyed alike.

Doors

Brand, color, model number, type (hollow core, solid or veneered), material (wood, metal, glass, plastic or other). Include hardware and hinge descriptions, fire rating.

Driveway & Finish

Type (asphalt, concrete, stamped concrete, pearock, pavers or other). For pavers, specify sand, concrete or asphalt underlay. If you live in a cold climate, inquire about heat strips under your driveway. Check for compaction and warranty information on all driveways. Find out if your driveway requires expansion joints.

Drywall

Type and thickness. 5/8" is generally used for ceilings and fire walls, veneered drywall or blue board for plastered areas and green board for tiled areas, such as bathrooms.

Electrical Fixtures

Brand, type, model number, finish.

Fascia	Material type and grade (wood, aluminum, vinyl, plastic, steel or other).
Fences	Type (chain link, wood, concrete, plastic or other), gates, hardware, lighting fixtures, finish.
Fireplaces	Brand, make and model number if preassembled. If custom made, specify damper, flue, lid, mantle, lintel, grout, surface material. Specify whether vent will be through wall or ceiling in either case.
Fixtures	Make, model number, color, brand, type, shade, weight.
Gates	Material (steel, pvc, wood, chain link, aluminum or other), hardware specifications, order number, sag resistance, finish.
Glass	Thickness, type (plate or tempered), shade, polished or raw cut, etched, beveled or sand blasted.
Glass Block	Brand, model number, style, size, silicone or mortar application.
Grout	Brand, color, shade, sanded or unsanded, cement or other material.
Hardware	Brand, model number, material type, color and finish.

Heating Systems	Brand, type, model number, size, heat recovery system.
Hinges	Brand, model number, material type, color and finish.
Insulation	Type, brand, thickness, R-value, faced or unfaced, square feet per package. R-value is the value of energy the insulation saves. Commonly used insulation is R-30 for ceilings, R-19 for floors, R-11 for interior walls and R-13 for exterior walls (may vary depending on the thickness of walls).
Mirrors	Thickness, quality, shading, beveling, etching or other. Brand, model if pre-cut.
Mouldings	Material (plastic, vinyl or wood). If wood, what type, thickness? Will the wood be clear (one length) or finger joint (random length pieces, joined together)? Will it be painted, stained or left natural?
Paint	Brand, color, quality, type (oil, acrylic or other), number of coats.
Paneling	Material type (wood, plastic, veneer, mica or other), color, model number and thickness.

Plaster	Brand, color, texture, thickness, type (veneer, brown coat). Specify mounting surface and whether bonding is required.
Plumbing Fixtures	Brand, type, model number, color, material (steel, aluminum, plastic, porcelain, glass, ceramic, stone, marble or other).
Plumbing Pipes	Material (plastic, copper or galvanized steel).
Primer	Brand, number, color, number of coats, concentrated or full strength.
Railings	Material, shape, finish and coats, size, type of millwork.
Roof Tile	Brand, color, type (Spanish "S", double barrel, mini-roll, large roll or other), material (clay, cement, metal, aluminum, steel, shingles or other).
Shelving	Material (wood, plastic, steel or other), color, amount of coats for finish, number of shelves, spacing, bracing materials. Brand, type and order number if pre-packaged.
Shingles	Make, model number, color, size, grade, thickness, weight, fire resistance rating,

	material (fiberglass, plastic, asphalt, rubber, metal, wood or other), manufacturer's specifications.
Shower Nozzles	Make, model number, type, size, style.
Siding	Material type (wood, plastic, veneer, mica or other), color, model number and thickness.
Sinks	Brand, make, model number, material (steel, fiberglass, porcelain, aluminum, plastic, cast iron or other), color, size and fixtures.
Stairs	Material (wood, steel, plastic or other), detailed finish, amount of risers, size of steps, assembly hardware.
Stucco	Synthetic or cement, pre-colored or gray, brand, thickness, number of coats.
Tar Paper	Number of layers, type. According to building code, generally 15 lb. felt for walls, 30 lb. felt for shingle roofs, 30 lb. and 90 lb. for tile roofs.
Tile	Brand, color, thickness, order number, type (Spanish, Italian, Mexican or other), material (marble, ceramic or other).

Toilets	Brand, make, model number, color, material (cast iron, fiberglass, wood or other), gallons of water used per flush.
Trim	Material (plastic, vinyl or wood). If wood, what type, thickness? Will the wood be clear (one length) or finger joint (random length pieces, joined together)? Will it be painted, stained or left natural?
Vanities	Brand, type, model number, color, size, material (mica, wood, plastic or other).
Wallpaper	Brand, batch and/or style code number, design, self-stick (prepasted) or paper-backed (glue required).
Water Systems	Brand, type, model number, filter type (charcoal, salt, potassium, reverse osmosis or other). Specify whole house or under sink unit(s).
Window Sills	Material (marble, wood, tile or other), size, color and thickness.
Windows	Brand, type, model number, frame material (wood, vinyl, aluminum or other), number of panes, insulated or uninsulated, glass description (tinting or shade), film tint or natural.

Window Treatments Make, model number, grade, material, size, trim, finish, hardware.

Wood Flooring Brand, order number, type, thickness, material (oak, pine, ash or other), solid, veneered, strip wood, parquet (wood tiles), self adhering or nail down, resin, heat-treated or other.

If you will be selecting items or materials on your own through purchase allowances, be sure the amounts allowed are sufficient. The best way to do this is by shopping around. Find out what quality you want for the items or materials you'll need. Get prices on your own *before you sign the contract*. For example, if you'd like carpeting that costs $15 per square yard, and you are offered an allowance of $11 per square yard, you'll know before you sign that the allowance for carpeting will be insufficient. This can help you avoid costly upgrades and change orders.

Specifications also apply to services and labor. These specifications, known as "general conditions" or "terms and conditions," include the terminology and clauses of the owner's requirements. Clean-up clauses, date of commencement, delay clauses, requests for license(s) and insurance documents, hours of work and other pertinent information should be listed here. The cost of permit fees is usually included. The contractor should apply for, and be the receiver of, all permits necessary to fully complete the project. However, owners should apply for any variances.

Contracts should include a clause stating that tool and equipment rentals, temporary services, transportation, trash removal and dump fees shall be the responsibility of the contractor. In most cases, these costs should be figured into the original contract price. They should not be billed separately to the owner as the job progresses.

Close attention to detail now will pay off later. If a contractor will need to move your furniture to remodel a room, be sure to state it in the contract. If you don't, you may end up paying extra. If you want to be there each time an inspector inspects your project, state it in the clauses. If you don't want workers consuming alcoholic beverages on your property, add a clause that restricts it. If you want to file your own Notice of Completion, put it in writing in the contract. *Put everything in writing.* Some clauses may be negotiable. Make a list of the items you would like to discuss with the contractor. Be sure to double check that what you agree upon is included in the final contract, word for word, item for item. Don't accept a verbal agreement of any kind.

The following pages provide an example of owner specifications.

Page ____ of ____

Owner's Specifications

Owner:

____

____

____

( ) ______________________

Project Address:

____

____

____

( ) ______________________

Architect:

____

____

____

( ) ______________________

Contractor:

____

____

____

( ) ______________________

Contract# ______________________

Date: ______________________

These specifications are hereby incorporated as part of the contract between owner and contractor. Owner shall provide contractor with a copy of specifications prior to contractor bidding the project.

1. **Contractor shall provide to owner (all information shall be typed or in writing and legible):**

- Copies of driver's license, contractor business license, worker's compensation, liability and property damage insurance, bond certificate and any and all other applicable licenses or registrations.
- List of references of recent and ongoing projects similar to owner's project.
- List of all suppliers and subcontractors who will be supplying materials and/or performing labor for project, including address, phone number and contact person.
- Copies of all subcontractors' license and insurance documents.
- Complete approved set of sealed plans prior to commencement.
- Schedule of hours when work will be performed.
- Releases of lien from all suppliers, subcontractors and contractor prior to receiving draw payments with final releases of lien prior to receiving final payment.
- Commencement and completion dates or Flow Chart showing project schedule.
- Warranty on all workmanship for a period of ______ year(s) following completion of project.
- All manufacturers warranties and/or guarantees.
- Physical samples of the following materials: ______________________

____

____

____

____

____

Page ______ of ______

2. **Contractor shall:**
- Apply for and pay for all applicable permits.
- Perform all work in accordance with local/state building codes.
- Notify owner of all scheduled inspection dates and type of inspection.
- Post permit and other applicable documents on job site under protective cover. Owner shall have access to these documents at all times.
- Provide amenities, including but not limited to, portable toilets, rental equipment, telephone service, food and beverages for himself, employees and subcontractors (or require each to provide their own).
- Notify owner of any changes in schedule and/or work through written change order in advance of said change. Change orders shall indicate the exact amount of time and cost for extras. Change orders shall become part of the original contract and shall include a place for both the owner(s) and contractor to sign upon approval.
- Notify owner of delivery dates of the following materials/supplies for owner's inspection prior to installation: ______________________

____

____

____

____

____

- Install materials in accordance with manufacturer's requirements.
- Include purchase allowances (for the cost of materials only) for the following:

Fixture(s)	Allowance amount

3. **Additional Requirements**
- All items specified in purchase allowances shall be selected by owner.
- Contractor shall purchase, deliver and install all items on purchase allowance list.
- A written receipt **on supplier's invoice** shall be presented to owner upon delivery of fixtures. In the event the amount of money spent on allowances exceeds the amount specified in the contract, the difference is due the contractor. If the amount spent is less than what is specified, the difference is due the owner.
- Owner reserves the right to purchase and deliver allowances personally and deduct the full specified amount listed above for the fixture(s). Contractor shall be responsible for installation of all fixtures.
- The following items shall be furnished by owner and installed by contractor:

____

____

Page ______ of ______

► All supplies or materials, whether new or used, which have been paid for by the owner, shall remain property of the owner unless otherwise specified and agreed to in writing.

► The following items shall be purchased and installed by contractor:

Item	Make	Model#	Color	Size

► Any substitutions shall be submitted by change order and approved by owner prior to purchasing the substituted item.

► All items, including fixtures and appliances, shall be in proper working order and satisfactory to owner prior to final payment.

Other conditions: ____

____

____

____

____

____

____

____

____

In the event of any discrepancies or disputes with regard to the plans or contract, the Owner's Specifications contained herein shall be valid and binding and override all others.

Acceptance

I HEREBY ACCEPT AND AGREE TO ALL OF THE TERMS LISTED HEREIN.

Owner	Date	Contractor	Date
Owner	Date	Witness or Notary Public	Date

NOTES:

NOTES:

Chapter Four

Selecting A Contractor: The Five Essentials

Many people rely on friends, acquaintances or business colleagues for referrals of contractors. This can be an effective way to obtain names of professionals to work with. It can provide you the opportunity to see completed work first hand and get an honest opinion about a contractor and his work.

Whether you meet a contractor through a referral from a friend or by any other means, do not be in a hurry to make a decision. If a contractor is highly recommended by an architect, find out whether they have a personal relationship or mutual business interests. In some cases, this can be unfavorable for the owner. Meet with a minimum of three contractors to discuss your project. Try to get a general impression of which contractor you would be able to communicate with best. Use professional courtesy and avoid confusion by scheduling each contractor to meet with you at a different time.

Many contractors offer free estimates. Confirm estimate charges, if any, with each contractor before they meet with you to discuss your project. **Don't be pressured into signing a contract during an initial visit.** Sometimes, estimates or proposals have contracts incorporated into them (*see Chapter Five*). As a general rule, estimates don't need an owner's signature. Don't be fooled by slick salesmen who tell you that you need to sign an estimate to acknowledge that it was given to you. You don't. *Be careful what you sign.*

When each contractor comes out for the initial visit, be prepared with a list of items to discuss. Present plans or drawings of the work and detailed specifications. Include any design ideas you have. Magazine clippings are sometimes helpful.

Be frank and to the point as you question each contractor's abilities and experience. A contractor is a consultant as well as an estimator. Try to obtain as much information as possible about your specific project. Take notes.

When interviewing contractors remember that they are interviewing you, too. Most reputable contractors will ask questions about your lifestyle, needs and expectations. Knowing whether an owner is interested in quality (not just price) can be a big factor in deciding whether or not to bid a job. A contractor is more likely to be interested in bidding a project when the owner has a complete, legible list of specifications and has researched costs. On the other hand, a contractor may *not* want to bid a project for an owner who does not have a specifications list and insists he will choose the lowest bidder.

The Five Essentials

There are five essentials to look for when selecting a contractor:

1) Fully licensed or registered (city, county and/or state, as required).

Ask each contractor to provide you with a copy of his license or registration. Check all documents for validity. *Be thorough.* Call the applicable authorities to confirm they are in effect and when they expire. Your city, town or county building department or State Department of Professional Regulation can confirm the documents are valid and when they expire. Ask to receive a copy of the confirmation in writing.

As an added precaution, ask the contractor to see his driver's license. Confirm that the name on the construction license, insurance and registration documents matches the name on the driver's license. Do not accept only a business card and the contractor's word. The contractor's construction license number may be printed on his card. In many states the law requires that construction license numbers be printed on **all** stationery and advertising. *It is extremely important to verify the number.* Anyone can have business cards printed.

2) Fully insured (worker's compensation, general liability, property damage).

Ask each contractor to provide you with copies of their insurance coverage. Also request copies of bond certificates, if bonded. Again, check the documents for validity and be thorough. Call the insurance agent or carrier to confirm the insurance is in effect and when it expires.

3) Clean complaint record.

Contact your local Better Business Bureaus, building associations and consumer protection agencies, as well as your State Department of Consumer Services and the Department of Professional Regulation. Also call the applicable licensing facilities with each contractor's license or registration number and name. Ask if each contractor has had any prior complaints filed against him or penalties imposed on his license. Inquire about fines, suspensions or other reprimands. Ask whether any problem situations are still unresolved.

4) Established local business, good reputation with local supply houses.

Confirm a physical address and local phone number for the contractor. Visit the address and call the phone number to be sure they are valid. Be wary of contractors offering P.O. box numbers or motel addresses as their place of business.

Check with local building supply houses the contractor uses regarding his reputation in the industry and ability to pay material bills. They may also be a source for references who have had work done by the contractor.

Request a financial statement from the contractor. You may also check a contractor's financial status through various credit sources.

5) Satisfied previous clients.

Request a list of references of previous clients. This is important whether a contractor is a referral from a friend, business colleague or anyone else. Choose referrals whose projects are similar to yours. Ask referrals about any recurring problems during or after construction. Ask specific questions regarding timeliness, response to concerns and requests, job clean up and overall job quality. Request to see the completed project. Your idea of a job well done may differ. Also ask about cost overruns, how well the contractor met deadlines and if he was on the job site on a regular basis.

Try to visit at least one project in progress. After meeting with three referrals, you should be able to make an educated decision on which contractor to use for your project.

Additional Suggestions

- Inquire about employees and subcontractors scheduled to work on your project. Ask the contractor to provide you with a list of names and copies of their licenses and insurance.

- Try to get a general impression of the contractor. Do you like him? Is he knowlcdgcable? Does he answer your questions directly? Does he appear to be organized and a good businessman?

- When visiting a project in progress, take note as to the actions of the contractor and his employees. Do they appear to be neat, orderly, courteous and professional?

- Ask the contractor how many projects he has in progress. Is he doing five projects at once? Are the projects small or extensive? Contractors who have several major projects going at one time may be hard to locate, or spend little time on your job site.

- Confirm that the contractor has a copy of all current local building codes.

Several associations offer informative consumer brochures on selecting a qualified contractor. Listed here are just a few:

The National Association of Home Builders Remodelors® Council offers a free eight page pamphlet with suggestions on what consumers should look for when interviewing and contracting with a remodeler. It's titled "How To Find A Remodeler Who's On The Level" and is available by sending a self-addressed, stamped #10 envelope to: NAHB Remodelors® Council, 1201 15th St., N.W., Washington, D.C. 20005.

The National Kitchen and Bath Association (NKBA) can provide you with a directory of its member contractors in your area qualified specifically in kitchen and bathroom remodeling. Write to them at: 687 Willow Grove Street, Hackettstown, NJ 07840, or call them toll-free at 1-800-FOR-NKBA.

The National Association of the Remodeling Industry (NARI) offers consumers a brochure titled "Select a Professional Remodeling Contractor," along with a list of selected members in their area. To receive this information call (800) 440-NARI. For more information about NARI call (703) 276-7600.

Contractor Regulatory Offices

Contractor and builder requirements vary from state to state and within each state. In some states, contractors need a license, while builders do not. Remember that local municipalities or counties may have their own requirements. Included here are the current state regulatory offices for contractors:

ALABAMA
State Licensing Board for General Contractors
400 South Union Street, Suite 235
Montgomery, Alabama 36130
(205) 242-2230

ALASKA
Department of Commerce & Economic Development
Division of Occupational Licensing
Contractor Section
P.O. Box 110806
Juneau, AK 99811-0806
(907) 465-2546 or (907) 465-3035

ARIZONA

Main Offices:
Registrar of Contractors
800 West Washington, 6th Floor
Phoenix, AZ 85007-2940
(602) 542-1525

Registrar of Contractors
400 West Congress, Suite 212
Tucson, AZ 85701-1311
(602) 628-6345

ARKANSAS

Contractors Licensing Board
621 East Capitol Avenue
Little Rock, AR 72202
(501) 372-4661

CALIFORNIA

Contractors State License Board
9835 Goethe Road
P.O. Box 26000
Sacramento,CA 95826
(916) 255-3900

COLORADO

Electrical and Plumbing Contractors:
State Board on Electrical & Plumbing
1390 Logan Street, Suite 400
Denver, CO 80203
(303) 894-2300

CONNECTICUT
Commissioner of Consumer Protection
Department of Occupational Licensing
State Office Building
165 Capital Avenue
Hartford, CT 06106
(203) 566-1810
(203) 566-3290

DELAWARE
Department of Finance
Division of Revenue
Carvel State Building
820 North French Street
P.O. Box 8911
Wilmington, DE 19899-8911
(800) 292-7826 (Delaware Only)
(302) 577-2554

DISTRICT OF COLUMBIA
Department of Consumer and Regulatory Affairs
614 H Street, N.W., Room 100
Washington, DC 20001
(202) 727-7089

FLORIDA
Department of Business & Professional Regulation
Construction Industry Licensing Board
7960 Arlington Expressway, Suite 300
Jacksonville, FL 32211-7467
(904) 359-6310

GEORGIA

Electrical, Plumbing, Utility and Air Conditioning Contractors:
Construction Industry Licensing Board
166 Pryor Street, S.W.
Atlanta, GA 30303-3465
(404) 656-3939

Out of State Contractors:
Bond may be required in addition to licenses and permits.
Contact:
Department of Revenue
270 Washington Street, S.W.
Room 311
Atlanta, GA 30334
(404) 656-4080

HAWAII

Department of Commerce & Consumer Affirs
Contractors License Board
1010 Richards Street
P.O. Box 3469
Honolulu, HI 96801
(808) 548-4100

IDAHO

No state regulatory office. Contact local municipalities.

ILLINOIS

No state regulatory office for residential contractors. Check local municipalities.

For Plumbing Contractors:
Office of Health Protection
Plumbing Section
525 West Jefferson Street
Springfield, IL 62761
(217) 524-0799

INDIANA

No state regulatory office for residential contractors.

For Plumbing Contractors:
Indiana Professional Licensing Agency
Plumber Regulation
302 West Washington Street
Room E-034
Indianapolis, IN 46204
(317) 232-2980

IOWA

Department of Employment Services
Division of Labor
1000 East Grand Avenue
Des Moines, IA 50319-0209
(515) 281-3606

KANSAS

Registration required through the Department of Revenue, however there is no state licensing office. Check local municipalities.

KENTUCKY

No regulatory office for residential contractors.

For Plumbing Contractors:
Department of Housing
Building and Construction
Division of Plumbing
1047 U.S. Highway 127 South, Suite 1
Frankfort, KY 40601-4337
(502) 564-3580

For Electrical Contractors:
State Fire Marshall
1047 U.S. Highway 127 South, Suite 1
Frankfort, KY 40601-4337
(502) 564-3626

LOUISIANA
Louisiana State Licensing Board for Contractors
P.O. Box 14419
Baton Rouge, LA 70898
(504) 765-2301

MAINE
No state regulatory office for residential contractors.

For Plumbing Contractors:
Plumbers Examining Board
State House Station 35
Augusta, ME 04333
(800) 541-5872
(207) 582-8723

For Electrical Contractors:
Electrician Examining Board
State House Station 35
Augusta, ME 04333
(800) 541-5872
(207) 582-8723

MARYLAND

For All Types of Contractors:
Comptroller of the Treasury
Retail Sales Tax Division
State License Bureau
301 West Preston Street
Baltimore, MD 21201-2383
(410) 225-1330

For Home Improvement Contractors:
Maryland Home Improvement Commission
Department of Licensing & Regulation
501 St. Paul Place, 8th Floor
Baltimore, MD 21202-2272
(410) 333-6309

MASSACHUSETTS

State Board of Building Regulations and Standards
Cashier's Office
One Ashburton Place, Room 1301
Boston, MA 02108
(617) 727-3200

MICHIGAN

Department of Licensing & Regulation
Residential Builders/Maintenance & Alteration
Contractors Board
Licensing Verification Unit
P.O. Box 30018
Lansing, MI 48909
(517) 373-0678

MINNESOTA

State of Minnesota
Department of Commerce
133 East 7th Street
St. Paul, MN 55101
(612) 296-2594

MISSISSIPPI

State Board of Contractors
2001 Airport Road, Suite 101
Jackson, MS 39208
(601) 354-6161

MISSOURI

No state regulatory offices for contractors. Check local municipalities.

MONTANA

Professional & Occupational Licensing
111 Jackson Street
Helena, MT 59620
(406) 444-4395

NEBRASKA

For All Contractors:
Nebraska Department of Revenue
P.O. Box 94818
Lincoln, NE 68509-4818
(402) 471-2971

For Electrical Contractors:
State Electrical Division
800 South 13th Street
Lincoln, NE 68508
(402) 471-3550

NEVADA

State Contractor Board
70 Linden Street
Reno, NV 89502
(702) 688-1141

State Contractor Board
1800 Industrial Road
Las Vegas, NV 89158
(702) 486-3500

NEW HAMPSHIRE

No regulatory office for residential contractors.

For Plumbing Contractors:
New Hampshire Plumbers Board
P.O. Box 1396
Concord, NH 03302
(603) 271-3267

For Electrical Contractors:
New Hampshire Electricians Board
P.O. Box 646
Concord, NH 03302
(603) 271-3748

NEW JERSEY

Plumbing & Electrical Contractors:
Department of Law & Public Safety
Division of Law
124 Halsey Street
P.O. Box 45029
Newark, NJ 07101
(201) 648-4010

Other Contractors:
Department of Banking
Division of Supervision
License Section
CN 040, 20 West State Street
Trenton, NJ 08625
(609) 292-5340 or (609) 292-5341

NEW MEXICO

Regulation and Licensing Department
Construction Industries Division
725 St. Michaels' Drive
P.O. Box 25101
Santa Fe, NM 87504
(505) 827-7030

NEW YORK
No state regulatory offices for contractors. Check local municipalities.

NORTH CAROLINA
Licensing Board for General Contractors
P.O. Box 17187
Raleigh, NC 27619
(919) 571-4183

NORTH DAKOTA
Secretary of State
600 East Boulevard Avenue
Bismarck, ND 58505-0500
(701) 224-3665 or (701) 224-2900

OHIO
No state regulatory offices for residential contractors. Check local municipalities.

OKLAHOMA
No state licensing of residential contractors.

Plumbing & Electrical Contractors:
Oklahoma State Department of Health
OLS-0203
1000 N.E. 10th Street
Oklahoma City, OK 73117-1299
(405) 271-5217

OREGON
Construction Contractors Board

700 Summer Street N.E., Suite 300
P.O. Box 14140
Salem, OR 97309-5052
(503) 378-4621

PENNSYLVANIA

No state regulatory offices for contractors. Check local municipalities.

RHODE ISLAND

Contractor's Registration Board
1 Capitol Hill
Providence, RI 02908
(401) 277-1268

SOUTH CAROLINA

Commercial Board:
State of South Carolina Licensing Board for Contractors
3600 Forest Drive
P.O. Box 11329
Columbia, SC 29211
(803) 734-4255

Residential Board:
South Carolina Department of
Labor, Licensing & Regulation
Residential Builders Commission
2221 Devine Street, Suite 530
Columbia, SC 29205
(803) 734-9051

SOUTH DAKOTA

No regulatory offices for residential contractors. Check local municipalities.

For Plumbing Contractors:
State Plumbing Commission
P.O. Box 807
Pierre, SD 57501
(605) 773-3429

TENNESSEE

Department of Commerce & Insurance
Board for Licensing Contractors
500 James Robertson Parkway, Suite 110
Nashville, TN 37243-1150
(800) 544-7693 (in TN) or (615) 741-8307

TEXAS

Texas Real Estate Commission
P.O. Box 12188
Austin, TX 78711-2188
(512) 465-3917

UTAH

Department of Commerce
Division of Occupational & Professional Licensing
160 East 300 South
P.O. Box 45805
Salt Lake City, UT 84145-0805
(801) 530-6532

VERMONT

No state regulatory office for residential contractors. Check local municipalities.

For Plumbing & Electrical Contractors:
Fire Prevention Division
National Life Building, Drawer 20
Montpelier, VT 05620
(802) 828-2107

VIRGINIA

Commonwealth of Virginia
Department of Professional & Occupational Regulation
3600 West Broad Street
Richmond, VA 23230-4917
(800) 552-3016 (in VA) or (804) 367-8511

WASHINGTON

Department of Labor & Industries
Contractor's Registration
P.O. Box 44450
Olympia, WA 98504
(206) 956-5226

WEST VIRGINIA

All Contractors:
Division of Labor
Contractor Licensing
Capitol Complex, Building 3, Room 319
Charleston, WV 25305
(304) 558-7890

For Electrical Contractors:
State Fire Commissioner
West Virginia State Fire Marshall
2100 Washington Street East
Charleston, WV 25305-0140
(304) 558-2191

WISCONSIN
Residential contractor registration will begin in 1995.

For Plumbing Contractors:
Department of Industry
Labor & Human Relations
Division of Safety and Building
P.O. Box 7969
Madison, WI 53707
(608) 266-3815

WYOMING
Labor Standards Division
Department of Employment
6101 Yellowstone
Room 259C
Cheyenne, WY 82002
(307) 777-7261

Worker's Compensation & Safety Division
Herschler Building
122 West 25th Street
2nd Floor, East Wing
Cheyenne, WY 82002
(307) 777-6760

NOTES:

NOTES:

Chapter Five

Understanding & Negotiating Contracts

In the building profession, the most common contract is a legal document between a contractor and the owner of a property. Within the contract, the contractor agrees to perform labor, supply necessary materials and/or subcontract required services to complete a project. The owner of the property agrees to make payment for the labor, materials and/or services.

There are many contract variations. Some contracts are for repairs or labor only. Many states require written contracts for building and improvements by law. Some states regulate contracts or have certain restrictions or provisions for specific contract amounts. Before hiring a contractor, check with your local building department or other applicable authorities.

To help you fully understand the contract process, let's first look at the stages that lead up to the signing of a contract — and how much they can vary.

Bids, Estimates, Proposals and Contracts

It is important to know the difference (and when there is no difference) between a bid, estimate, proposal and contract. Bids and estimates should be on a standard form or company letterhead. Every bid or estimate should include detailed specifications of the project and a total price for all services, labor and materials. In addition, bids and estimates should be signed and dated **by the contractor** issuing them. *Bids and estimates generally do not need or require an owner's signature. Be careful what you sign.* **Bids and estimates signed by the owner may act as a contract agreement.** Many contractors offer free estimates. Confirm estimate or bid charges, if any, with each contractor before they meet with you to discuss your project. Keep a copy of each bid to compare with the contract before signing.

Proposals and contracts are written agreements for work to be done and materials to be supplied. They should include a place for the date and both the contractor's and owner's signatures. Proposals and contracts are often one and the same (many proposals have a contract incorporated into them). Some contractors use proposal forms to submit an estimate. Others use proposal forms as the actual contract. When issued separately, their content and specifications should be checked and rechecked to be sure they are identical.

All bids, estimates, proposals and contracts should be figured and signed by **the contractor who will actually perform the work**. Never sign anything immediately. Carefully read all paperwork before signing. Completely read everything over and fully understand and agree with what you are signing for. Don't be pressured or intimidated by anyone who tells you their bid is only good for one week, or that they won't offer the same price again after they leave. Most reputable contractors don't use pressure sales tactics and will afford you the time you need to thoroughly check out a contract.

Comparing Bids

Most contractors strive to submit a bid with the best price, as opposed to the cheapest price. To determine the best price, a contractor must consider the full scope of a project. He must carefully figure in all costs from beginning to end, and then add a reasonable profit.

Bids from low to high can vary thousands of dollars. There is much controversy about which bid area is best — low, middle or high. In the construction industry it is the general consensus that middle to upper-middle bids are often the best choice. These bids usually allow for labor and material cost increases and other unforeseen situations, helping to prevent the contractor from cutting corners to finish the job.

Bids that are too low are a red flag for a host of problems. These can range from low wages for workers to contracts that are purposely bid low. Sometimes numerous items are left out of a contract and need to be added through change orders. This can be a very costly process, and in the end may cost you more than the highest bid (not to mention the stress, aggravation and delays). However, extremely low bids given by a highly reputable contractor may warrant asking if something was left out or somehow misunderstood.

Being vague will cause you to gather bids that are not comparable. Being specific (colors, model numbers, brand names, purchase allowances, etc.) will give you more accurate bids and allow the contractors to bid "apples for apples". Make certain you provide *each* contractor with the same final list of specifications and purchase allowances or model numbers.

Ask each contractor to supply you with product samples or brochures with their written estimate. Hold on to the product samples you are given and label them with the name of the contractor who provided them. They will be useful to compare with the samples given to you by each of the other contractors, or samples you've obtained on your own. Once you have selected a contractor, you can use the samples to compare with the actual products installed during your project.

Time and Materials Bids

Time and materials contracts can be risky due to the time factor, which may encourage the work to be done slowly. Also, it may be difficult to track actual labor time. Homeowners sometimes request time and materials bids when doing part of the work themselves.

Payments for this type of contract are generally made as stages of the project are completed. Original receipts (from the material supply company), signed time sheets and releases of lien should be in the owner's possession before making payments. Time and materials bids or contracts should include a maximum cost amount (cap). Be sure to include a complete list of specifications and purchase allowances for materials.

Labor Only Bids

Labor contracts are good for small projects, although they are usually between subcontractors and general contractors. This type of contract may be acceptable if you can coordinate the purchase and delivery of materials. *If you enter into a direct contract with a laborer, make sure that your liability insurance covers you for the laborer's activities.* Confirm in writing the amount of time the project will take and the rate per man hour. Specify a maximum cost amount. If you are supplying materials for a labor only contract, be sure to order a sufficient amount to complete each stage of the project. Also make sure the supplies arrive on time. Running out of supplies and/or not having supplies delivered on time can cause excess labor charges even though no work is being completed.

In General

Some projects are quicker and easier to bid than others. These generally involve window replacement, roofing, siding, driveways and slabs, or other jobs bid on a cost-per-unit basis. For example, roofing and siding are bid in units of 100 square feet, called "squares" in the industry. Concrete is poured by the cubic yard and finished by the square foot. Tile and wood flooring is installed by the square foot, while carpeting is installed by the square yard.

Additions and remodeling projects that require that the structure of a home be modified will take longer to bid. This is due to more extensive figuring and the obtaining of bids from subcontractors, when necessary. Also, the contractor should figure in a price for moving, protecting or securing existing property susceptible to damage as a result of the work being done.

The size of a building company may also affect the bid amount. This is due to profit and overhead factors. A smaller company where the contractor does most of the work himself, with his own employees, may offer a lower bid. Or a large company, which buys materials in quantity, may offer a lower bid.

Consumers will often obtain two or three bids from reputable contractors and still have a big difference in price ranges. The most common reason: the consumers are not specific enough to allow the contractors to bid "apples for apples." Stating the brand name and color alone is not sufficient. Adding exact model numbers, style numbers and other specifications will help you get more precise bids.

Rules to Remember:

- Request that the contractor who will actually do the work come out and personally bid your job. This often helps reduce misunderstandings.
- All bids should be in writing. NEVER accept a verbal bid under any circumstances.
- Do not accept bids over the phone for jobs that haven't been seen in person.
- Bids too good to be true usually are.

- Be sure each bid specifies which materials the contractor is to use and what requirements they meet.

- Read and clarify all information, conditions and specifications provided by the contractor or any other parties (such as subcontractors) included in a bid. If you are unsure about any portion of the contract, seek professional advice. Be sure to obtain copies of all bids, estimates and proposals.

- Be sure all bids include your list of specifications.

- Bids are more likely to be competitive if you provide each contractor with your list of specifications and purchase allowances. In instances where all bids submitted are over budget, obtain recommendations from each contractor on how to cut project costs. After you have decided on the most suitable changes, resubmit your new specifications to all of the contractors for new bids.

The Contract

Once you have reviewed all submitted bids and decided which contractor you would like to use, the next step is to secure a written contract. Many people accept a contractor's written contract. However, some owners prepare their own contract and have an attorney review it before presenting it to a contractor. Others have an attorney draw up a contract for them to present to a contractor. Both options are becoming common. A real estate attorney with construction experience, who specializes in construction law, may be the best source for these type contracts.

Keep in mind that some contractors and builders will not accept owner contracts. Be sure to confirm this information before calling a contractor or builder out to look at your project.

The following is a checklist of information that should be included in a contract agreement:

✓

___ **Owner's Name/Address/Phone Number**
List the full name of all owners of the property. Also include a complete address, with zip code, so that you can receive Notices to Owner and other important documents. If possible, provide more than one phone number for use in case of an emergency.

___ **Job Name/Address/Phone Number**
In many instances the information you will list here will not be the same information listed under owner's name, address and phone. The property may be a second home, vacant land or office site. If a numerical address and street number are not available, (such as sometimes occurs with vacant land) include the complete legal description.

___ **Contractor's Company Name/Address/ Phone Number/Fax Number/Officers**
Obtain the company's full legal name. If the company is a corporation, the business name should include Inc., Incorporated or Corporation. Also get

a complete address (not a P.O. box), along with telephone and fax numbers for each partner or corporate officer.

___ **Contractor Name/Address/Phone Number/ License or Registration Number/ Copies of Construction License, Insurance and Bond Documents/Copy of Driver's License**

The contract should include the contractor's full name, address (not a P.O. box) and day and evening phone numbers. Include a pager number, if available. Obtain copies of all city, town, county and state licenses. Also get copies of insurance and bond documents. Don't forget a copy of the contractor's driver's license.

___ **Total Contract Amount**

This is the final price of the contract, including all work to be completed and materials to be supplied.

___ **Payment Terms or Draw Schedule**

Most established contractors are financially capable of starting a project with a deposit of 5% or less. Many contractors have accounts with local suppliers who extend credit for 30 days or more. This can allow enough time for the contractor to progress a project to the first draw or stage. Keep in mind that the fee for a building permit can be paid directly to the building department. If you want to do this, be sure to state it in the contract. Use discretion when making any deposit payments. If you are unsure, consult your attorney.

When a contractor submits a proposal, the draws should be scheduled as needed with a retainer (generally 5-20%) upon completion of the project. If, for instance, the contract amount is $100,000, six draws should be adequate. Draw amounts will vary according to type, duration and cost of a project. Some states regulate both the amount of deposit and the payment schedule through home improvement laws.

Be sure a minimum 5-20% retainer is held on the final draw until final inspection and satisfactory completion of all items noted on punch lists. (Punch lists contain items that need to be completed. They are usually compiled by the owner, or owner and architect.) Many owners request a retainer for a period of 2-4 weeks after project completion. This provides time to confirm that everything is in working order and completed to specifications. It's important to check the not-so-obvious, such as heating systems if you're building in the summer, or cooling systems if you're building in the winter. *Carefully check all roof repairs.*

If your project is being financed through a lender, confirm in writing that the payments will be made directly to the owner, not the contractor. Another possibility may be to have the lender make the checks out in the name of the contractor *and* the owner. That way you can regulate the monies being paid out according to the work completed. It also gives you the opportunity to check invoices for duplicates (so you won't pay for anything twice) or supplies that were never received. Many owners have found they were charged for supplies delivered to another job site.

Some lenders make payments to builders who present them with receipts for materials used on a project. Other lenders pay after their own inspection of the project. *Don't rely on a lender to make inspections before they release draw payments.*

Inspections performed by building authorities and inspections performed by lenders are different. Inspections performed by building authorities are done to enforce code requirements. Inspections performed by lenders are often done only to confirm that a portion of work has been completed. Neither guarantee quality work. The "inspection" performed by a lender could be made by a loan officer with no construction experience at all. How does the loan officer confirm what has been completed? In many cases, by taking pictures. Working with a loan officer with extensive construction experience may offer some peace of mind. Controlling the monies and inspecting the project yourself (or with an independent inspector) can be helpful too. If the lender is unwilling to make the checks out in your name, consider asking them to pay the subcontractors and suppliers directly.

Be sure to specify in writing that the lender must receive all lien releases from subcontractors, suppliers and the contractor upon each payment. Also require the lender to provide you with a copy of each lien release. Some lenders stamp the back of every draw payment check with a "lien release" stamp. By signing the check, the receiver acknowledges that there are no liens being placed against the property for materials and/or services they have provided.

Funding control companies are another option for the disbursement of money during a project. These companies act as a "middle man" between the owner and contractor, subcontractors or suppliers when payments are due. Rather than making payments directly to the contractor, subcontractors or supply companies, the owner authorizes the funding control company to make the payments.

Most funding control companies provide you with special forms or vouchers to fill out each time you authorize a payment. The forms or vouchers should not be given to, or filled out by, anyone other than the owner authorizing payment. These forms or vouchers should only be given out when payments are due according to the draw schedule specified in the contract. *Be sure the work has been completed correctly.*

When selecting a funding control company, confirm whether it is licensed and how long the company has been in business. Make sure the company specializes if funding control for your type of project (residential or commercial). Like lender and building department inspections, funding control company inpections are not guarantees of quality work. Most funding control companies perform inspections to verify receipt of materials and completion of work only. Once again, you should confirm in writing the experience (preferably *construction* experience) of the inspector who will be inspecting your project.

If necessary, get a professional opinion as to whether the amount of money requested per draw payment will be equal to the amount of work completed to date.

Thoroughly inspect the entire project before signing a certificate of completion and making the final payment. **All** final releases of lien should be in writing and in your possession **before** you release the final draw. This includes the contractor's final release of lien and final invoice showing that the contract is paid in full.

____ **Change Order and Extras Clause**

First, *make sure the contract states that the owner has the right to make changes to the project after the contract is signed.* When making a change or adding materials to a project, the details should always be specified in writing and signed and dated by all parties involved. Two types of forms, called "Change Orders" or "Extras to Contract Agreements," are generally used.

Change orders are written descriptions of proposed changes to a contract. They should include prices and the amount of time needed for completion. All change orders should become a part of the original contract. Both the owner and contractor should sign change orders before their production. Careful, proper planning of your project can eliminate the need for extras or change orders.

Common reasons for change orders include the desire to change colors, textures or material brands. The cost for change orders that are a direct result of an error by the contractor or architect should not be the owner's responsibility. Rarely, new environmental requirements may develop which cause the need for a change order. Check with the proper local authorities about these requirements before authorizing any changes.

Any changes verbally expressed, whether in person or by phone, should be put in writing and signed by all parties involved. Do not assume that writing a note and presenting it to the contractor (or giving it to one of his crew) will serve as a written change order.

Change orders for materials or appliances may include restocking fees. Confirm these fees before requesting a change order. *Obtain copies of all change orders or extras to contract agreements.*

___ **Dates of Acceptance/Commencement/Completion (Project Schedule)**

The date the contract is signed is the date of acceptance. The commencement date is the date the actual work begins (after the permit is issued). *The commencement date can sometimes be months after the acceptance date.* Blueprint processing problems, supply shortages and contractors doing too many projects at once are common reasons for delayed commencement dates.

Ask each contractor bidding on your project to confirm the exact date, in writing, when he is available to start work. Ask about other jobs each contractor is currently working on. It's important that the contractor you choose is readily available to commence and supervise your project on the commencement date listed on the contract.

If you've hired an architect, inform the contractor that you will schedule a meeting between the owner, architect and contractor prior to signing the contract. This is an important part of your project planning. Having two professionals consulting about your project with you prior to commencement can save you time, money and aggravation. Architects are design professionals, while most contractors are concerned with the physical development of projects, including the cost of materials and labor. Coordinating their services will help keep your project on schedule. Ask for a flow chart.

Your project may require interior or landscape design, or other design services. Be sure to consult with all tradespersons who will be performing design work before your draftsman completes your plans. You may want designers to attend the meeting between homeowner, contractor and architect. This is especially helpful in instances where the design work will be extensive or heavily detailed. It may help you avoid delays in your project schedule.

___ **List of Subcontractors**
This list should include subcontractors for plumbing, electrical, heating/air conditioning, plastering, roofing and any other subcontractor services scheduled. Be sure to obtain phone numbers, addresses, insurance certificates and license and/or registration numbers.

___ **List of Suppliers**
The contract should include the name, address and phone number of all suppliers who will be sending materials or appliances to the project. This list is useful for determining who will be sending Notices to Owner and/or lien releases.

___ **Project Delay Clause**
Discuss the estimated duration of project with the contractor. Request a date of completion in the written contract. A clause may be added for penalties when a project is completed after the completion date. Penalties may then be assessed, as long as

the reason for late completion is not an exception (weather, acts of public authority, etc.) in the contract. Remember that labor or material shortages and strikes can also cause project delays. Keep in mind that if you request a penalty for late completion, the contractor may request a bonus for early completion.

___ **Lien Release Requirements**

Be sure the contract states that you will be released from all liens. Make certain to obtain **all** releases of lien with each payment. Many lenders stamp each check issued to a contractor or supplier with a lien release. Once signed, the check becomes a release of lien. If your project will be financed through a lender, get an agreement in writing stating that you will receive copies of all releases of lien upon receipt.

___ **Cancellation Clause**

This clause should state the owner's right to cancel the contract in the event of poor workmanship, inability or refusal to meet contract requirements and other reasons. Cancellation clauses are sometimes used when a contract is contingent upon the owner being able to obtain financing.

___ **Warranties and/or Guarantees**

Be sure all warranties and guarantees (including their duration) of both materials and workmanship are stated in the contract (*see Chapter Eight*).

___ **Performance Statement**
The contract should include a clause stating that work performed and supplies used will meet all specifications and building code requirements.

___ **Unforeseeable Work Clause**
These clauses allow for remedies of unforeseeable work, such as leaking pipes or electrical wiring problems behind walls. The clause should state the owner's right to prior notice before unforeseen work is done. It should also include the right to obtain other bids for the additional work, in the event the contractor's price is high.

___ **Substitution of Materials Clause**
This clause should state that a contractor may not substitute any materials or appliances without the written consent of the owner.

___ **Financing Information**
If applicable, include all financing terms, lock-in rates and schedules. Also see *Cancellation Clause.*

___ **Permit Requirements**
The contract should specify the contractor as the applicant and receiver of all required permits for the project, *not the owner.* If the contract doesn't state either one, do not assume that the contractor is the receiver. Get everything in writing.

The contract must include the price for permits, whether paid as part of the contract price or as a purchase allowance. Purchase allowances are sometimes used to purchase permits when the amount due is unknown until the plans are processed. When using purchase allowances for permits, the contractor must provide *original receipts on a building department letterhead or invoice.*

___ **Liability Clause**
Liability clauses are statements of the contractor's responsibility for the entire project, including damages, injuries and warranties. These clauses should be carefully reviewed by an attorney.

___ **Arbitration Clause**
Contracts should include a clause stating how disputes will be resolved if they arise. Consider consulting with an attorney regarding arbitration.

___ **Clean up Clause**
Make sure a clean up clause is included in the contract. Clean ups should be done at regular intervals throughout the project, with a final clean up done upon completion.

___ **Work Schedule**
Be sure to get a flow chart or written schedule of the days weekly and hours per day when work will be in progress. For example, 8 am - 5 pm, Monday through Saturday.

___ **Purchase Allowances**
Contracts with purchase allowances included in the

price allow you to shop for materials of your own choice. Purchase allowances are usually for the purchase of materials only and should not include any labor costs. For example, when doing a bathroom remodeling project, you may request allowances for tile, electrical, plumbing and bath fixtures, mirrors, wallpaper, cabinets and shower doors.

If the amount of money paid for the allowances exceeds the amount specified in the contract, the amount in excess is due to the contractor. Accordingly, if the amount of money paid for the allowances is less than the amount specified in the contract, the customer should receive the difference as a refund. This is provided there are no other arrangements in writing. Be sure your purchase allowances are for a reasonable amount or you may get stuck with extras or poor quality materials. Always obtain **original** receipts from suppliers.

Contractors usually prefer to purchase allowance items with their supply accounts. This enables them to keep the project on schedule and coordinate measurements of fixtures being selected.

If you decide to choose your own fixtures and let the contractor do the purchasing, he may charge an additional handling fee for purchase, pickup and delivery. Many contractors offer their contractor discounts to their clients. Therefore, after deducting any handling fees, you may find you are still getting the best price on the fixtures.

___ **Owner-Supplied Items**
Include a list of items the owner will purchase and deliver to the project for the contractor to install.

___ **Owner-Performed Labor**
List here the projects the owner will personally perform labor for, or will hire a laborer to complete, separate from the contract with the contractor.

___ **Surplus Materials Clause**
If remodeling, you should include a surplus materials clause listing any materials or fixtures you want to keep. The contractor should remove any remaining surplus materials before final payment.

___ **Amenities Clause**
This clause should state that the homeowner is not responsible for supplying water, ice, portable toilets, rental equipment, telephone service or other amenities for the contractor, his employees or subcontractors.

___ **Insurance Clause**
This clause should state the amount and types of insurance required to be maintained by the contractor.

___ **Specifications**
As outlined in Chapter Three, owner specifications should include all supplies to be used, all services to be provided and all terms and conditions of the contract.

In Conclusion

Before signing the contract, be sure all of the information on paper is to your expectations. If anything is unclear, call the contractor and architect and have them verify what it means. If you are still unclear about any part of the contract, you may want to consult an attorney.

NEVER sign a blank or incomplete contract. Each page should be labeled with its corresponding number (i.e., page 1 of 4, 2 of 4). Your specifications and general conditions should be included and signed by both parties. Cross through all blank spaces that do not pertain to your job. Obtain an **original** of the contract, with signatures, for your records.

A place for the owner and contractor to sign and date as acceptance of the agreement is generally located on the last page of the contract. The signing of the contract should be witnessed by one or two other persons or a notary.

NOTES:

Chapter Six

The Lien Law: What You Should Know

What It Is

In plain terms, the construction lien law allows unpaid persons who have provided labor and/or building materials for a project to make legal claim for payment against an owner's property.

How You Can Protect Yourself

It is important to become educated about and understand the lien law. Lien law provisions vary from state to state. To confirm lien law provisions in your state, contact the Division of Consumer Services or the Real Estate Commission for your area. You can also check your local courthouse library. Look for state statutes on construction law, real estate liens and mechanic's liens.

The following will also help you to protect yourself:

1) Be certain a Notice of Commencement (or equivalent document, depending on state law) has been filed before the start of your remodeling or construction project. The Notice of Commencement form is provided where building permits are issued for your area. This form must be recorded with the Clerk of the Circuit Court in the county where the

property is undergoing construction or remodeling. A certified copy should be posted at the job site (or an affidavit stating it has been recorded with a copy of the Notice of Commencement attached). Make sure all of the information listed on the Notice of Commencement is correct. The Notice of Commencement identifies:

A) property owner
B) location of project
C) contractor
D) description of project
E) lender
F) surety
G) amount of bond
H) date of commencement

Failing to record a Notice of Commencement, or obtaining one with incorrect information, may add to the possibility of having to pay twice for labor or materials. It may also prevent you from passing required code inspections.

2) **Never** make a payment without receiving a release of lien from the contractor, suppliers and subcontractors, as applicable.

Notices to Owner

Contractors or suppliers will usually send you a Notice to Owner form after they are contracted to supply materials or labor for your project. It is not a lien on your property, however, it is a legal document advising you that they are supplying materials and/or labor for your project. Be advised that not all states require contractors to send Notice to Owner documents. Releases of lien should be obtained from any firm or individual who sends you a Notice to Owner once they have been paid. Compare the list of individuals or companies who have provided you with a Notice to Owner to the list of subcontractors and suppliers provided to you by your contractor.

Responsibility for Obtaining Releases of Lien

You may want to stipulate in the contract that the contractor is responsible for providing you with all releases of lien. If this clause is not included, it may become your responsibility.

How a Lien is Imposed Against Property

Let's say, for example, that you hire a contractor for a $20,000 kitchen remodeling project. You pay the contractor the $20,000 in full, but he neglects to pay three material suppliers who supplied materials used on your project. There is no clause in your contract stating that the contractor is responsible for providing you with releases of lien. The contractor has neglected to supply you with the releases. Each supplier who has not been paid may then place a lien against your property until they have been paid.

Legal Effects of Liens

You cannot sell or transfer ownership of your property until all outstanding liens have been paid. You may also find you are unable to refinance your mortgage or secure a home equity loan if you have any outstanding liens.

In rare instances, property owners with outstanding liens may be required to sell their property to satisfy them.

Liens Placed Without Just Cause

If you feel a lien has been placed against your property without just cause, ask your local court for a Notice of Contest of Lien. Also consider consulting an attorney. Lienholders who receive a Notice of Contest of Lien are issued a set time period in which to either sue you or cancel the lien.

Types of Lien Releases

Partial Releases of Lien

As each payment is made for a particular part of your job, you should receive a partial release of lien for all materials and labor to date. Upon completion of the entire project, make sure you receive a notarized final release of lien.

Unconditional Releases of Lien

Unconditional releases of lien are those that are free and clear of any contingencies. By signing them, the supplier of materials or labor acknowledges that he has been paid in full. Some "unconditional" releases state that unresolved claims for extras and change orders are not included.

Conditional Releases of Lien

Conditional releases of lien are those that are contingent upon a specified action. Many conditional releases are contingent upon checks being paid out by the bank they are drawn on.

In General

It is important to include a release of lien clause in your contract. This clause should state that the contractor must furnish you with a **partial** release of lien from himself, all subcontractors and suppliers *upon each payment*. It should also state that a **final** release of lien shall be received from all subcontractors, suppliers and thc contractor *upon completion and final payment*. Remember not to make any payments without receiving lien releases.

The following page shows an example of a final release of lien.

FINAL RELEASE OF LIEN

KNOWN ALL MEN BY THESE PRESENTS, that the undersigned, ______________________________, for the sum of ________________(______) lawful money of the United States of America, to the undersigned in hand paid, the receipt whereof is hereby acknowledged, does hereby waive, release, remise and relinquish the undersigned's right to claim, demand, impress, or impose a lien or liens in the sum of $________________ for work done or materials furnished (or any other kind or class of lien whatsoever) up to the ______ day of ____________, 19___ on the property described on Exhibit "A" below.

EXHIBIT A

____

____

____

____
COMPANY NAME

BY:____________________

____
TITLE (President or V.P.
or Owner or Partner)

Original Contract Date:____________________
This payment reflects final payment for contract and no balance due to the contractor and all subcontractors hired by the contractor.

Signed, sealed and delivered
in the presence of:

____
(Witness)

____
(Witness)

Reference:
Check Number: __________
Check Date: __________

NOTES:

Chapter Seven

Insurance & Bonding: Coverage and Requirements

It's essential to confirm that both you and your contractor have proper insurance coverage before your construction project begins. There are several types of insurance that affect construction. Always verify coverage, limits and expiration dates. Be sure to obtain a copy of every policy, including any addendums.

Contractor Liability Insurance Coverage

Before signing a contract, you should ask the contractor for a copy of his general liability insurance policy. Thoroughly check the limits of liability. Policies for $100,000 and $300,000 in coverage are common. For an additional fee (which is sometimes surprisingly minimal) the coverage can be increased to $500,000 or $1,000,000. If you feel the amount is not sufficient, and you want the extra coverage, consider asking the contractor to increase his policy limit.

Several states don't require contractors to carry liability insurance. *It's important to make sure the contractor you choose has liability coverage, whether required by state law or not.* If the contractor doesn't carry liability coverage, and someone other than an employee is injured on your property during construction, you are likely to be liable. In many instances homeowners who are sued for injuries must resort to the use of their homeowner's insurance policy for coverage of medical bills.

Confirm the date of expiration on the policy. If the policy will expire during your project, add a clause stating that the contractor must show proof of new coverage *as of the date of expiration*. This helps prevent the coverage from lapsing and helps ensure that it will be renewed. No further payments should be made until proof of renewal is received.

Types of Contractor Liability Coverage

General Liability generally covers bodily injury and property damage.

Professional Liability covers lawsuits for misrepresentation or false information resulting in damage or injury.

Commercial Vehicle Liability, also known as **Business Auto Liability**, provides coverage for company-owned vehicles in the event of an accident.

Non-Owned Vehicle Liability provides coverage for employee-owned vehicles being used for work purposes.

Umbrella Liability is excess liability coverage that is added to an existing policy.

Product Liability covers accidental injury or damage to persons or personal property caused by a product provided by the insured.

Completed Operations or Tail Liability provides coverage for physical injury *after* the project is completed, but as a result of work performed.

Other Coverage

Workers' Compensation provides bodily injury coverage for employees injured while at work.

Builder's Risk is generally required by lenders from builders. It covers damage to a structure and damage or theft of materials on a project.

Homeowner Insurance Coverage

Homeowners' insurance policies usually protect homes from loss or damage due to theft, fire, liability, acts of nature and other events. Most all-inclusive policies cover the structure and contents of a home as well as personal liability, in the event of injuries to others.

When having renovations or remodeling done, it is important to confirm your coverage and limits. Does your homeowner's policy cover worker's compensation and extended liability for workers? If you will be doing part of the work yourself, consider what type of coverage you need in case you become injured. Some contractors require homeowners to carry property damage and liability insurance for the amount of their contract.

Ask about temporary policies covering loss, damage or theft of materials and supplies on the project, if not covered by the contractor.

Appraisals

Remodeling or adding a room to a home will require a change in insurance. In either case, a new appraisal should be made. It is important to know the difference between replacement value, market value and purchase price. Replacement value is the value of an item without depreciation factored in. Market value is how much an item is currently worth in the market. Purchase price is, of course, the actual amount you paid. In order to receive sufficient money for replacement with brand new items, coverage should be for full replacement value.

According to many appraisal experts, reappraisals should be done every five to ten years. The frequency depends on market changes and improvements to the home. In areas where home values are rising rapidly, appraisals should be done more often than in neighborhoods where the home values barely fluctuate.

Many items cannot be replaced for the dollar amount allowable under a standard insurance policy. For example, an antique stained-glass window is not likely to be replaced for the same price as a standard casement window. Therefore, these items should be appraised at full replacement value.

Listed here are some of the items that should receive careful attention during an appraisal:

- Unique handcrafted work
- Expensive and/or exotic woods
- Specially designed glass (including stained-glass windows and hand etched glass)
- Imported tile and fixtures
- Custom built items
- Antique items and fixtures
- Rare marble
- Hand-carved paneling and banisters
- Commercial appliances

Home Offices

Home offices can sometimes present particular insurance problems. Some companies refuse to write policies that include home office coverage. Others will, with certain limitations. Be sure to check available coverage with your present insurance company *before* signing a contract.

Title Insurance

Title insurance may have construction restrictions. Check your policy. Extended coverage policies are available through many title companies. This extended coverage sometimes includes lien protection as well. Some owners purchase the extended coverage themselves, while others require the contractor to pay for the coverage. The average cost for extended coverage is $100 - $125.

Bonds and Bonding

Although bonds are issued by insurance companies and regulated by insurance commissioners, they are not a form of insurance. This is a common misconception. The function of a bond is to serve as a credit guarantee that a contractor will perform what is specified by the bond. Contractors must qualify to become bonded. To qualify, a contractor must submit an application along with business and personal insurance and financial information.

By law, many states require contractor bonds. Various municipalities require contractors to post a bond if they are not in business locally. There are many different types of bonds. These are some of the more common ones:

License and Permit Bonds are required by some states and municipalities from licensed contractors. Their purpose is to ensure proper business ethics on the contractor's part. At the same time, they protect the licensor(s) from claims as a result of providing the license and/or permit.

Performance Bonds guarantee the completion of a project according to the contract and plans, including all terms, conditions and specifications.

Payment Bonds can provide protection from liens for materials and labor that could be filed against your property. However, even when a payment bond has been issued, you should still obtain all lien releases.

Contract Performance and Payment Bonds are usually combinations of Performance Bonds and Payment Bonds.

Contract Bonds provide a guarantee that a contractor will fully perform all provisions of a contract.

Maintenance Bonds are issued to guarantee quality of workmanship and materials for a specified period of time (usually one year or more).

General Information

Keep in mind that bonding companies only pay up to the face value of the bond. Therefore, bonds should be purchased with face value amounts equivalent to or greater than the cost of the project.

Obtain copies of all bonds, just as you would the contract, insurance paperwork, lien releases and other documents. *Be sure the bond number is legible.*

Bond amounts are often refunded (less any applicable fees) back to the party who purchased the bond upon receipt of a signed Notice of Completion. Authorization from the owner may also be required. It is a good idea to file your own Notice of Completion. That way you will have the opportunity to ensure that everything has been done to your specifications before the contractor is paid (or paid back) in full.

NOTES:

Chapter Eight

Getting The Most Out Of Warranties & Service Contracts

In order to get the most benefit from warranties and service contracts, it's important to understand how they work. Before making a purchase, carefully read the warranty or service contract to find out what kind of protection it offers. By researching the kinds of warranties and service contracts that are available, you can make the best choice for your needs.

Product Warranties

Federal Law On Product Warranties

The Magnuson-Moss Warranty Act of 1975, as amended, is a federal law that regulates warranties for consumer products. Its stipulations require that warranties offered for consumer products costing more than $10 state whether they are full or limited. For consumer products costing over $15, the warrantor is required to provide specific information regarding coverage. This information must be presented in a form that can be easily read. The warrantor must also make warranty information available to consumers *before* they buy a consumer product costing over $15. The Magnuson-Moss act applies only to written warranties, not services.

Types of Product Warranties

There are several types of product warranties. They include verbal, written, implied, full, limited, multiple and express warranties. Here is a brief overview of each type:

Verbal Warranties are oral representations made by a seller or member of a sales staff. *All verbal warranties should be put in writing before a purchase is made.*

Written Warranties are given with many products. It is important to confirm what is covered and what is omitted from coverage. While some warranties cover both parts and labor, others exclude certain parts and types of repairs. Many warranties cover parts, but require you to pay labor fees. Confirm how long the warranty is valid and whether the seller or manufacturer provides service and repairs. Ask if the warranty is transferable, and if so, whether there is a fee. Also confirm the company's policy for product failure. Some companies will repair or replace an item, while others may offer a full refund. Check for any restrictions before you buy a product (or request your contractor to buy it). *Carefully read warranties that act as a rider to your homeowner's insurance.*

Implied Warranties are a part of state law. All states have implied warranties. Nearly every purchase made carries an implied warranty. The exceptions are products labeled for sale "as is," or that clearly state in writing there is no warranty being given. Many states do not allow "as is" sales.

There are two types of implied warranties: *warranties of merchantability* and *warranties of fitness for a particular use*. A "warranty of merchantability" guarantees that a product will do what it is meant to do, such as a washing machine being able to wash. A "warranty of fitness for a particular use" applies when a seller recommends and states that a product is suitable for a specific type of use.

Implied warranties on consumer products may be in effect for as long as four years. To confirm your state's implied warranty coverage for consumer products, contact your local consumer protection office or an attorney.

Full Warranties give a buyer several benefits. They provide warranty service at no cost, and often include postage or shipping costs to return repaired items. Full warranties are transferable to new owners when a product is sold. They do not change or limit the length of time of an implied warranty. The company will offer replacement or a full refund when unable to repair a product. And there are no requirements for receiving service, such as filling out a dated registration card. These features make full warranties attractive to buyers.

Limited Warranties usually restrict one or more of the benefits offered by full warranties. They also have one loophole: the duration of a limited warranty can have an impact on the implied warranty duration. When a seller provides a limited written warranty, the law allows the seller to make a provision restricting the implied warranty to the same length of time. An example might be a stove with a two-year limited warranty. Although its implied warranty may normally be for four years, the two-year limited warranty would only allow the implied warranty to be effective for two years.

Multiple Warranties are combinations of full and limited warranties. A multiple warranty may offer full coverage for specified parts of a consumer product, while offering limited coverage for other parts of the same product. These warranties are also used when full coverage is valid only during a specified time within the warranty period (i.e., the first six months of use).

Express Warranties are guarantees or promises made by a seller at the time of a sale. They can be issued in the form of a verbal commitment, written into advertising or presented to the buyer as a formal certificate. Remember that the Magnuson-Moss Warranty Act only covers consumer products with *written* warranties.

Product Warranties In General

Research product warranties *before purchasing*. **Obtain** product warranties *before installation* so that you can confirm the installation is being done according to the manufacturer's specifications. Certain types of installation will void warranties. If you are unsure about a certain form of installation, call the manufacturer and ask. If you cannot be there during installation of each warrantied item, consider having your architect or another qualified professional confirm that the work has been done according to the manufacturer's specifications. This may take some schedule coordination on the part of the contractor and the inspecting professional. This is especially important when the installation of a product is under concrete, inside of a wall or on a roof.

Improper maintenance can also affect and/or void a warranty. Most appliances and materials come with proper maintenance instructions and directions. Using a product for something other than it was intended will often void a warranty.

Obtain receipts from the actual supplier whenever possible. Save all product receipts for proof of the purchase date. They can also be used to show that you were the original purchaser.

Be sure that all warranties include the complete name, address and telephone number of both the seller (distributor) and the manufacturer.

Additional Information

To learn more about warranties, write for a free copy of *Warranties* to: Public Reference, Federal Trade Commission, Washington, D.C. 20580

Disputes

Any correspondence sent to the seller or manufacturer regarding problems or disputes should be sent by certified mail. If you are unable to resolve a dispute, contact your local consumer protection agency for assistance or consider using an arbitrator.

Some warranties require an attempt at arbitration before a case can be taken to court. For more information on dispute resolution, write to the FTC and ask for: *How to Resolve Consumer Disputes* and *Road to Resolution: Handling Consumer Disputes*," Public Reference, Federal Trade Commission, Washington, D.C. 20580

Extended Product Warranties

Although many companies state that they offer "extended warranties," they are actually offering service contracts. The difference: A warranty is included in the price of a product. Service contracts are purchased separately, at an extra cost to the buyer.

Before purchasing a service contract, confirm whether you already have the same coverage under the warranty. If the coverage is comparable, consider waiting until your warranty expires before purchasing a service contract. *Make sure the supplier will allow you to purchase a service contract at a later date.*

Get specifics of what's included in a service contract. Check which parts are covered and for how long. If the product has no likelihood of needing repairs, or the cost for repairs would be low, you may want to bypass buying a service contract. Selecting items known for quality workmanship and buying from reputable companies is important. For more information on buying service contracts, write to: *Service Contracts*, Public Reference, Federal Trade Commission, Washington, D.C. 20580.

Construction Warranties

Before hiring a contractor, be sure their warranty policy on workmanship and materials is clearly written out in the contract. One year should be the minimum. Try to negotiate warranties for longer than one year, especially on foundations and framing work. Five years should be the minimum on roofs. Confirm when the warranty starts (the date of completion, the date of installation, or the date of occupancy). Since the warranty is likely to be "limited," confirm exactly what the limited clauses omit.

Implied Warranties In Construction

Almost every state has implied warranties which affect construction. These are warranties that automatically go into effect when there is no written warranty. Although their components may vary from state to state, the concept is generally the same: the contractor warrants that the work was done properly with quality materials. Implied warranties for construction can be from one to ten years, depending on state law. Beware of clauses in a contract that can affect or restrict an implied warranty. Restrictions on implied warranties are not allowed in several states. To confirm your state's implied warranty coverage, contact your state consumer protection agency (*see Chapter Nine for listings*) or consult your attorney.

Long-Term Home Warranties

In the construction industry, service contracts are often referred to as long-term warranties. Before buying a long-term warranty from a contractor you should first consider the following factors:

The Company's Age

How long a company has been in business, *performing the same type of work*, is the first consideration. Buying a long-term warranty from a contractor who has only been in business for nine months and has had only two clients during that time may be risky. Why? Two reasons. First, a high percentage of contractors go out of business in their first year. If a contractor is not around to honor a warranty, what good will it be? Secondly, if the contractor has only done work for two other clients, what indicator will you have of how well he backs his warranty?

A reputable contractor who has been in business a number of years may be more likely to stay in business for the duration of your warranty. Some contractors offer a longer warranty to enhance their company's reputation for customer service.

The Company's Occupation

As mentioned above, it is important to confirm that a company has been doing the same type of work (the longer, the better) they are offering a warranty for. If a company offers you a contract with an extended warranty for a new roof, and for the past ten years they have exclusively done plumbing work, be careful. Although the names may remain the same, the occupations of companies often change.

The Contractor's References

Before hiring a contractor, ask for the names and numbers of at least five references who have purchased his extended warranty. Ask for clients who have had work done similar to yours. If available, ask for the names of clients whose warranty with the contractor has been in effect for a number of years. Contact the references and ask about any problems that have occurred and how the contractor handled them. Ask how specific problems were taken care of and how long it took the contractor to come out and fix them. If possible, visit the actual projects and talk to the references in person. Preferably, ask the contractor to refer clients who are not close friends or relatives.

Deductibles

Watch out for deductibles on warranties. Just as there are deductibles on most insurance policies, some construction warranties don't go into effect until you've spent a designated dollar amount.

New Home Warranties

There are several companies that offer ten-year new home warranties, also referred to as "homeowner protection plans." These warranty programs have attracted an abundance of attention, including that of Congress. Due to numerous consumer complaints, Congress began formal hearing procedures to investigate the practices of these companies. If you decide to purchase a ten-year new home warranty plan, be sure to check how long the company has been in business. Also ask for references and check the company's complaint history.

Many consumers feel there is an alternative to ten-year warranty plans: hire a reputable contractor known for quality work and make sure the project is done correctly from the start.

NOTES:

Chapter Nine

Your Consumer Rights: What They Are & Where To Go For Help

Whenever consumers purchase products or hire an individual or company for services, they have various consumer rights. Building and remodeling is no different. In most states, after signing a building or remodeling contract, the following applies:

1) You may cancel contracts signed at a location other than the seller's regular business address. The exceptions are cases where you have specifically requested the goods and/or the services. Cancellation must be made by midnight of the third business day after the signed transaction.

2) You may cancel contracts you sign with a door to door solicitor, or contracts which will be paid in installments for more than 90 days, so long as they are canceled by midnight of the third business day after the signed transaction.

3) In the instance of an emergency home repair or repairs, where the owner specifically requests emergency service, the three-day policy is not allowable.

In many states, contractors are required by law to inform you of your cancellation rights. They must also provide you with a notice of cancellation form.

Should you decide to cancel your contract within the three-day time period, be sure to send your cancellation notice by certified mail, return receipt requested. Keep a file with copies of all correspondence.

Before you hire a contractor, check to see if complaints have been filed against him. Find out whether there are still unresolved cases, and how long the company has been in business. To confirm this information, contact your local Better Business Bureau and/or state, county or local consumer protection agencies. These organizations can also offer assistance and answer questions regarding consumer rights.

The following pages contain a national listing of consumer protection agencies by state.

Alabama

Ms. Priscilla Black Duncan
Consumer Affairs Division
Office of Attorney General
11 South Union Street
Montgomery, AL 36130
(205) 242-7334
(800) 392-5658
(toll free in AL)

Alaska

The Consumer Protection Section in the Office of Attorney General has been closed. Consumers with complaints are being referred to the Better Business Bureau, small claims court and private attorneys.

American Samoa

Ms. Jennifer Joneson
Assistant Attorney General
Consumer Protection Bureau
P.O. Box 7
Pago, Pago, AS 96799
011 (684) 633-4163
011 (684) 633-4164

Arizona

Ms. H. Leslie Hall
Chief Counsel
Consumer Protection
Office of Attorney General
1275 West Washington Street
Suite 259
Phoenix, AZ 85007
(602) 542-3702
(602) 542-5763
(consumer info and complaints)
(800) 352-8431
(toll free in AZ)

Ms. Noreen Matts
Assistant Attorney General
Consumer Protection
Office of Attorney General
402 W. Congress Street
Suite 315
Tucson, AZ 85701
(602) 628-6504

Arkansas

Ms. Kay DeWitt, Director
Consumer Protection Division
Office of Attorney General
200 Tower Building
323 Center Street
Little Rock, AR 72201
(501) 682-2341 (voice/TDD)
(800) 482-8982
(toll free voice/TDD in AR)

California

Dr. C. Lance Barnett
Interim Director
California Department of Consumer Affairs
400 R Street, Suite 1040
Sacramento, CA 95814
(916) 522-1700 (TDD)
(800) 344-9940
(toll free in CA)

Office of Attorney General
Public Inquiry Unit
P.O. Box 944255
Sacramento, CA 94244-2550
(916) 322-3360
(800) 952-5225
(toll free in CA)
(800) 952-5548
(toll free TDD in CA)

Colorado

Consumer Protection Unit
Office of Attorney General
1525 Sherman Street, 5th Floor
Denver, CO 80203
(303) 866-5189

Connecticut

Ms. Gloria Schaffer
Commissioner
Dept. of Consumer Protection
165 Capitol Avenue
Hartford, CT 06106
(203) 566-2534
(800) 842-2649
(toll free in CT)

Mr. Robert M. Langer
Assistant Attorney General
Antitrust/Consumer Protection
Office of Attorney General
110 Sherman Street
Hartford, CT 06105
(203) 566-5374

Delaware

Ms. Mary McDonough
Director
Division of Consumer Affairs
Department of Community Affairs
820 North French Street
4th Floor
Wilmington, DE 19801
(302) 577-3250

Mr. Stuart Drowos
Deputy Attorney General for
Economic Crime & Consumer
Protection
Office of Attorney General
820 North French Street
Wilmington, DE 19801
(302) 577-2500

District of Columbia

Mr. Hampton Cross
Acting Director
Department of Consumer and
Regulatory Affairs
614 H Street, N.W.
Washington, DC 20001
(202) 727-7120

Florida

Karen K. MacFarland, Director
Department of Agriculture and
Consumer Services
Division of Consumer Services
407 South Calhoun Street
Mayo Building, Second Floor
Tallahassee, FL 32399-0800
(904) 488-2221
(800) 435-7352
(toll free information
and education in FL)

Mr. Jack A. Norris, Jr., Chief
Consumer Litigation Section
4000 Hollywood Boulevard
Suite 505 South
Hollywood, FL 33021
(305) 985-4780

Ms. Mona Fandel, Bureau Chief
Consumer Division
Office of Attorney General
4000 Hollywood Boulevard
Suite 505 South
Hollywood, FL 33021
(305) 985-4780

Georgia

Mr. Barry W. Reid
Administrator
Governors Office of
Consumer Affairs
2 Martin Luther King, Jr. Dr., S.E.

Plaza Level - East Tower
Atlanta, GA 30334
(404) 651-8600
(404) 656-3790
(800) 869-1123
(toll free in GA)

Hawaii

Mr. Philip Doi, Director
Office of Consumer Protection
Department of Commerce and
Consumer Affairs
828 Fort St. Mall, Suite 600B
P.O. Box 3767
Honolulu, HI 96812-3767
(808) 586-2636

Mr. Gene Murayama
Investigator
Office of Consumer Protection
Department of Commerce and
Consumer Affairs
75 Aupuni Street
Hilo, HI 96720
(808) 933-4433

Mr. Glenn Ikemoto
Investigator
Office of Consumer Protection
Department of Commerce and
Consumer Affairs
3060 Eiwa Street
Lihue, HI 96766
(808) 241-3365

Ms. Pamela LaVarre
Investigator
Office of Consumer Protection
Department of Commerce and
Consumer Affairs
54 High Street
Wailuku, HI 96793
(808) 243 5387

Idaho

Mr. Brett De Lange
Deputy Attorney General
Office of Attorney General
Consumer Protection Unit
Statehouse, Room 119
Boise, ID 83720-1000
(208) 334-2424
(800) 432-3545 (toll free in ID)

Illinois

Governors Office of
Citizens Assistance
222 South College
Springfield, IL 62706
(217) 782-0244
(800) 642-3112 (toll free in IL)
(Only handles problems related to
state government)

Ms. Patricia Kelly, Chief
Consumer Protection Division
Office of Attorney General
100 West Randolph, 12th Floor
Chicago, IL 60601
(312) 814-3000
(312) 793-2852 (TDD)

Ms. Kathryn Patterson, Director
Department of Citizen Advocacy
100 West Randolph, 13th Floor
Chicago, IL 60601
(312) 814-3289
(312) 814-3374 (TDD)

Indiana

Mrs. Lisa Hayes, Chief Counsel
Consumer Protection Division
Office of Attorney General
219 State House
Indianapolis, IN 46204
(317) 232-6330
(800) 382-5516 (toll free in IN)

Iowa

Mr. Steve St. Clair
Assistant Attorney General
Consumer Protection Division
Office of Attorney General
1300 East Walnut Street, 2nd FL
Des Moines, IA 50319
(515) 281-5926

Kansas

Ms. Theresa Marcel Nuckolls
Deputy Attorney General
Consumer Protection Division
Office of Attorney General
301 West 10th
Kansas Judicial Center
Topeka, KS 66612-1597
(913) 296-3751
(800) 432-2310 (toll free in KS)

Kentucky

Mr. Robert V. Bullock, Director
Consumer Protection Division
Office of Attorney General
209 Saint Clair Street
Frankfort, KY 40601-1875
(502) 564-2200
(800) 432-9257 (toll free in KY)

Mr. Robert Winlock, Administrator
Consumer Protection Division
Office of Attorney General
107 South 4th Street
Louisville, KY 40202
(502) 595-3262
(800) 432-9257 (toll free in KY)

Louisiana

Ms. Tamera R. Velasquez, Chief
Consumer Protection Section
Office of Attorney General
State Capitol Building
P.O. Box 94095
Baton Rouge, LA 70804-9095
(504) 342-9638

Maine

Mr. William N. Lund
Superintendent
Bureau of Consumer Credit
Protection
State House Station No. 35
Augusta, ME 04333-0035
(207) 582-8718
(800) 332-8529 (toll free)

Mr. Stephen Wessler, Chief
Consumer and Antitrust Division
Office of Attorney General
State House Station No. 6
Augusta, ME 04333
(207) 626-8849 (9 a.m.-1 p.m.)

Maryland

Mr. William Leibovici, Chief
Consumer Protection Division
Office of Attorney General
200 St. Paul Place
Baltimore, MD 21202-2021
(410) 528-8662 (9 a.m.-3 p.m.)
(410) 565-0451 (DC Metro area)
(410) 576-6372
(TDD in Baltimore area)
(800) 969-5766 (toll free)

Ms. Emalu Myer
Consumer Affairs Specialist
Eastern Shore Branch Office
Consumer Protection Division
Office of Attorney General
201 Baptist Street, Suite 30
Salisbury, MD 21801-4976
(301) 543-6620

Mr. Larry Munson, Director
Western Maryland Branch Office

Consumer Protection Division
Office of Attorney General
138 East Antietam Street
Suite 210
Hagerstown, MD 21740-5684
(301) 791-4780

Massachusetts

Mr. Robert Sherman, Chief
Consumer Protection Division
Department of Attorney General
1 Ashburton Place
Boston, MA 02103
(617) 727-8400
(information & referral to local consumer offices that work in conjunction with the Department of Attorney General)

Ms. Gloria Cordes Larson
Secretary
Executive Office of Consumer Affairs and Business Regulation
One Ashburton Place, Room 1411
Boston, MA 02108
(617) 727-7780
(information & referral only)

Mr. Thomas J. McCormick
Assistant Attorney General
Western Massachusetts Consumer Protection Division
Department of Attorney General
436 Dwight Street
Springfield, MA 01103
(413) 784-1240

Michigan

Mr. Frederick H. Hoffecker
Assistant in Charge
Consumer Protection Division
Office of Attorney General
P.O. Box 30213
Lansing, MI 48909
(517) 373-1140

Minnesota

Mr. Curt Loewe, Director
Consumer Services Division
Office of Attorney General
1400 NCL Tower
St. Paul, MN 55101
(612) 296-3353

Mississippi

Ms. Leslie Staehle
Special Assistant Attorney General
Dir., Office of Consumer Protection
P.O. Box 22947
Jackson, MS 39225-2947
(601) 354-6018

Mr. Joe B. Hardy, Director
Bureau of Regulatory Services
Department of Agriculture and Commerce
500 Greymount Avenue
P.O. Box 1609
Jackson, MS 39215-1609
(601) 354-7063

Missouri

Office of Attorney General
Division of Consumer Protection
P.O. Box 899
Jefferson City, MO 65102
(314) 751-3321
(800) 392-8222 (toll free in MO)

Mr. Henry Herschel, Chief Counsel
Consumer Protection Division
Office of Attorney General
P.O. Box 899
Jefferson City, MO 65102
(314) 751-3321
(800) 392-8222 (toll free in MO)

Montana

Consumer Affairs Unit
Department of Commerce
1424 Ninth Avenue
Box 200501
Helena, MT 59620-0501
(406) 444-4312

Nebraska

Mr. Paul N. Potadle
Assistant Attorney General
Consumer Protection Division
Department of Justice
2115 State Capitol
P.O. Box 98920
Lincoln, NE 68509
(402) 471-2682

Nevada

Mr. John P. Kuminecz
Commissioner of Consumer Affairs
Department of Commerce
State Mail Room Complex
Las Vegas, NV 89158
(702) 486-7355
(800) 992-0900
(toll free in NV)

Mr. Ray Trease
Consumer Services Officer
Consumer Affairs Division
Department of Commerce
4600 Kietzke Lane, B-113
Reno, NV 89502
(702) 688-1800
(800) 992-0900 (toll free in NV)

New Hampshire

Chief
Consumer Protection and
Antitrust Bureau
Office of Attorney General
State House Annex
Concord, NH 03301
(603) 271-3641

New Jersey

Ms. Emma N. Byrne
Director
Division of Consumer Affairs
P.O. Box 45027
124 Halsey Street, 7th Floor
Newark, NJ 07101
(201) 648-6534

Ms. Lauren F. Carlton
Deputy Attorney General
New Jersey Division of Law
P.O. Box 45029
124 Halsey Street, 5th Floor
Newark, NJ 07101
(201) 648-7579

New Mexico

Consumer Protection Division
Office of Attorney General
P.O. Drawer 1508
Santa Fe, NM 87504
(505) 827-6060
(800) 678-1508 (toll free in NM)

New York

Ms. Rachel Kretser
Assistant Attorney General
Bureau of Consumer Frauds
and Protection
Office of Attorney General
State Capitol
Albany, NY 12224
(518) 474-5481

Mr. Richard M. Kessel
Chairperson and Executive Director
New York State Consumer
Protection Board
99 Washington Avenue

Albany, NY 12210-2891
(518) 474-8583

Mr. Richard M. Kessel
Chairperson and Executive Director
New York State Consumer
Protection Board
250 Broadway, 17th Floor
New York, NY 10007-2593
(212) 417-4908 (complaints)
(212) 417-4482 (main office)

North Carolina

Mr. Alan S. Hirsch
Special Deputy Attorney General
Consumer Protection Section
Office of Attorney General
Raney Building
P.O. Box 629
Raleigh, NC 27602
(919) 733-7741

North Dakota

Ms. Heidi Heitkamp
Office of Attorney General
600 East Boulevard
Bismarck, ND 58505
(701) 224-2210
(800) 472-2600 (toll free in ND)

Mr. Tom Engelhardt, Director
Consumer Fraud Section
Office of Attorney General
600 East Boulevard
Bismarck, ND 58505
(701) 224-3404
(800) 472-2600 (toll free in ND)

Ohio

Mr. Mark T. D'Allessandro
Consumer Frauds/Crimes Section
Office of Attorney General
30 East Broad Street
State Office Tower, 25th Floor
Columbus, OH 43266-0410
(614) 466-4986 (complaints)
(614) 466-1393 (TDD)
(800) 282-0515 (toll free in OH)

Mr. William A. Spratley
Office of Consumers' Counsel
77 South High Street, 15th Floor
Columbus, OH 43266-0550
(614) 466-9605 (voice/TDD)
(800) 282-9448 (toll free in OH)

Oklahoma

Ms. Jane Wheeler
Assistant Attorney General
Office of Attorney General
Consumer Protection Division
4545 No. Lincoln Blvd., Suite 260
Oklahoma City, OK 73105
(405) 521-4274

Mr. John L. McClure, Administrator
Department of Consumer Credit
4545 No. Lincoln Blvd., Suite 104
Oklahoma City, OK 73105-3408
(405) 521-3653

Oregon

Mr. Terry Leggert
Attorney in Charge
Financial Fraud Section
Department of Justice
1162 Court Street, NE
Salem, OR 97310
(503) 378-4732

Pennsylvania

Mr. Renardo Hicks, Director
Bureau of Consumer Protection
Office of Attorney General
Strawberry Square, 14th Floor
Harrisburg, PA 17120

(717) 787-9707
(800) 441-2555 (toll free in PA)

Mr. Michael Butler
Deputy Attorney General
Bureau of Consumer Protection
Office of Attorney General
1251 South Cedar Crest Blvd.
Suite 309
Allentown, PA 18103
(215) 821-6690

Mr. Daniel R. Goodemote
Deputy Attorney General
Bureau of Consumer Protection
Office of Attorney General
919 State Street, Room 203
Erie, PA 16501
(814) 871-4371

Mr. Robin David Bleecher
Attorney in Charge
Bureau of Consumer Protection
Office of Attorney General
132 Kline Village
Harrisburg, PA 17104
(717) 787-7109
(800) 441-2555 (toll free in PA)

Mr. E. Barry Creany
Deputy Attorney General
Bureau of Consumer Protection
Office of Attorney General
Professional Building
P.O. Box 716
Edensburg, PA 15931
(814) 949-7900

Mr. John E. Kelly
Deputy Attorney General
Bureau of Consumer Protection
Office of Attorney General
21 South 12th Street, 2nd Floor
Philadelphia, PA 19107
(215) 560-2414
(800) 441-2555 (toll free in PA)

Ms. Stephanie L. Royal
Deputy Attorney General
Office of Attorney General
Manor Complex, 5th Floor
564 Forbes Avenue
Pittsburgh, PA 15219
(412) 565-5394

Mr. J.P. McGowan
Deputy Attorney General
Bureau of Consumer Protection
Office of Attorney General
214 Samters Building
101 Penn Avenue
Scranton, PA 18503-2025
(717) 963-4913

Rhode Island

Ms. Christine S. Jabour, Esquire
Consumer Protection Division
Department of Attorney General
72 Pine Street
Providence, RI 02903
(401) 274-4400
(401) 274-4400 ext. 2354 (TDD)
(800) 852-7776
(toll free in RI)

Mr. Edwin P. Palumbo
Executive Director
Rhode Island Consumers' Council
365 Broadway
Providence, RI 02909
(401) 277-2764

South Carolina

Mr. Ken Moore
Assistant Attorney General
Consumer Fraud and Antitrust
Section

Office of Attorney General
P.O. Box 11549
Columbia, SC 29211
(803) 734-3970

Mr. Steve Hamm, Administrator
Department of Consumer Affairs
P.O. Box 5757
Columbia, SC 29250-5757
(803) 734-9452
(803) 734-9455 (TDD)
(800) 922-1594 (toll free in SC)

South Dakota

Mr. Jeff Hallen
Assistant Attorney General
Division of Consumer Affairs
Office of Attorney General
500 East Capitol
State Capitol Building
Pierre, SD 57501-5070
(605) 773-4400

Tennessee

Mr. Steven Hart
Deputy Attorney General
Division of Consumer Protection
Office of Attorney General
450 James Robertson Parkway
Nashville, TN 37243-0485
(615) 741-3491

Ms. Elizabeth Owen, Director
Division of Consumer Affairs
500 James Robertson Parkway
Nashville, TN 37243-0600
(615) 741-4737
(800) 342-8385 (toll free in TN)

Texas

Mr. Joe Crews
Asst. Attorney General and Chief
Consumer Protection Division
Office of Attorney General
P.O. Box 12548
Austin, TX 78711
(512) 463-2070

Mr. Robert E. Reyna
Assistant Attorney General
Consumer Protection Division
Office of Attorney General
714 Jackson Street, Suite 800
Dallas, TX 75202-4506
(214) 742-8944

Ms. Valli Jo Acosta
Assistant Attorney General
Consumer Protection Division
Office of Attorney General
6090 Surety Drive, Room 113
El Paso, TX 79905
(915) 772-9476

Mr. Richard Tomlinson
Assistant Attorney General
Consumer Protection Division
Office of Attorney General
1019 Congress Street, Suite 1550
Houston, TX 77002-1702
(713) 223-5886

Mr. Stephen C. McIntyre
Assistant Attorney General
Consumer Protection Division
Office of Attorney General
1208 14th Street, Suite 900
Lubbock, TX 79401-3997
(806) 747-5238

Mr. Michael Winget-Hernandez
Assistant Attorney General
Consumer Protection Division
Office of Attorney General
3201 North McColl Rd., Suite B
McAllen, TX 78501
(512) 682-4547

Mr. Aaron Valenzuela
Assistant Attorney General
Consumer Protection Division
Office of Attorney General
115 East Travis Street, #925
San Antonio, TX 78205-1607
(512) 225-4191

Utah

Ms. Francine A. Giani, Director
Division of Consumer Protection
Department of Commerce
160 East 3rd South
P.O. Box 45804
Salt Lake City, UT 84145-0804
(801) 530-6001

Vermont

Ms. Marilyn S. Skoglund
Assistant Attorney General
and Chief
Public Protection Division
Office of Attorney General
109 State Street
Montpelier, VT 05609-1001
(802) 828-3171

Virgin Islands

Mr. Clement Magras, Comm.
Department of Licensing and
Consumer Affairs
Property & Procurement Bldg.
Subbase #1, Room 205
St. Thomas, VI 00802
(809) 774-3130

Virginia

Mr. Frank Seales, Jr., Chief
Consumer Litigation Section
Office of Attorney General
Supreme Court Building
101 North Eighth Street
Richmond, VA 23219
(804) 786-2116
(800) 451-1525 (toll free in VA)

Ms. Betty Blakemore, Director
Division of Consumer Affairs
Department of Agriculture and
Consumer Services
Room 101, Washington Building
P.O. Box 1163
Richmond, VA 23209
(804) 786-2042

Washington

Ms. Renee Olbricht, Investigator
Consumer Protection Division
Office of Attorney General
P.O. Box 40118
Olympia, WA 98504-0118
(206) 753-6210

Ms. Sally Sterling
Director of Consumer Services
Consumer and Business
Fair Practices Division
Office of Attorney General
900 Fourth Avenue, Suite 2000
Seattle, WA 98164
(206) 464-6684
(800) 551-4636 (toll free in WA)

Mr. Owen Clarke, Chief
Consumer and Business
Fair Practices Division
Office of Attorney General
West 1116 Riverside Avenue
Spokane, WA 99201
(509) 456-3123

Ms. Cynthia Lanphear
Consumer and Business
Fair Practices Division
Office of Attorney General
1019 Pacific Avenue

3rd Floor
Tacoma, WA 98402-4411
(206) 593-2904

West Virginia

Mr. Donald L. Darling, Director
Consumer Protection Division
Office of Attorney General
812 Quarrier Street, 6th Floor
Charleston, WV 25301
(304) 558-8986
(800) 368-8808
(toll free in WV)

Wisconsin

Mr. John Alberts, Administrator
Division of Trade and Consumer Protection
801 West Badger Road
P.O. Box 8911
Madison, WI 53708
(608) 266-9836
(800) 422-7128
(toll free in WI)

Ms. Margaret Quaid
Regional Supervisor
Division of Trade and Consumer Protection
927 Loring Street
Altoona, WI 54720
(715) 839-3848
(800) 422-7128
(toll free in WI)

Regional Supervisor
Division of Trade and Consumer Protection
200 No. Jefferson St., Suite 146A
Green Bay, WI 54301
(414) 448-5111
(800) 422-7128 (toll free in WI)

Regional Supervisor
Consumer Protection Office
3333 North Mayfair Rd., Suite 114
Milwaukee, WI 53222-3288
(414) 266-1231

Mr. James D. Jeffries
Assistant Attorney General
Office of Consumer Protection and Citizen Advocacy
Department of Justice
P.O. Box 7856
Madison, WI 53707-7856
(608) 266-1852
(800) 362-8189 (toll free)

Mr. Nadim Sahar
Assistant Attorney General
Office of Consumer Protection
Department of Justice
Milwaukee State Office Bldg.
819 North 6th Street, Room 520
Milwaukee, WI 53203-1678
(414) 227-4949
(800) 362-8189 (toll free)

Wyoming

Mr. Mark Moran
Assistant Attorney General
Office of Attorney General
123 State Capitol Building
Cheyenne, WY 82002
(307) 777-7874

NOTES:

Chapter Ten

Common Questions From Consumers

Most consumers have numerous questions when they are about to hire a contractor. Getting the right answers is important, because education is a homeowner's best defense against problems. The following questions are the most commonly asked by consumers:

Q.

Where can I find a reputable contractor?

A.

Many people feel that finding a reputable contractor is like looking for a pot of gold at the end of a rainbow. A suggestion might be to drive around local rural or suburban areas near you and visit job sites similar to your own. If you like what you see (neatness, organized workers, etc.), write down the information listed on the contractor's sign or truck. You might also write down the address or street number. Then call the local building department with the information and ask for the name of the contractor doing the project. Building departments can be a good source for reputable contractors. Building industry associations and organizations are other referral sources. Be sure to check all referral's credentials thoroughly.

Q.

The contractor we would like to hire presented us with a contract with the arbitration clauses crossed out. Now we're not sure we want to use him. Any suggestions?

A.

Yes. Present the contractor with *your* list of specifications and general conditions (including arbitration) and negotiate along those terms. Never accept a contract with standard items or conditions crossed out. Consider consulting an attorney.

Q.

The contractor we would like to use provided us with a copy of his liability insurance. We called the agent and were told his policy is up for renewal. What should we do?

A.

Put a clause in your contract which states that the contractor must maintain all applicable insurance policies for the duration of your project. Consider adding a stipulation requiring the contractor to provide proof of renewal **as of the day of cancellation** of the previous policy before any further payments will be made.

Q.

Is it okay to use an out-of-state contractor for a local project?

A.

Yes, if the contractor has obtained all local licenses and insurance and meets any other requirements so that he may legally pull permits. Consider the following before hiring an out-of-state contractor:

1) How will you see previous jobs done by an out-of-state contractor?

2) Will the contractor provide you with names and numbers of previous clients as references?

3) Can the contractor provide you with the names and numbers of previous suppliers? Does the contractor have local supply accounts established? Will the contractor use local subcontractors on your project?

4) Are there any complaints or unresolved issues listed with the Better Business Bureau in the state the contractor previously worked in?

5) Is the contractor familiar with local building codes?

6) How will the contractor back his warranty?

Q.

The contractor working on my home has requested a draw payment and has run out of release of lien forms. What should I do?

A.

Do not make the payment until you get the release. Lien release forms are available from most office supply stores (you may want to keep a few blank forms on hand).

Q.

I signed a contract for a remodeling project and paid a deposit to the contractor. Now I want to cancel. Any suggestions?

A.

Several points should be considered. The first consideration is when the contract was signed. In some states, if the contract was signed within three days of your decision to cancel and you did not specifically request the materials and/or services, you can probably cancel the contract.

If it is past the three-day period, you may want to approach the contractor and request cancellation. If you signed a contract, then were informed that your company was transferring you to another state in two weeks, you may have a valid reason for requesting a refund. However, if the contractor turned down other projects to commence yours, you may have to forfeit all or part of your deposit. Depending on how many hours he has put in, you may owe him additional money.

Another consideration is where the contract was signed. If it was signed at the contractor's place of business, you may not be able to cancel. Consult your attorney.

Q.

My brother's friend does carpentry work. Is it okay to use him for my home addition?

A.

This is a personal decision. Generally speaking, it is not recommended to use the services of family or friends for home improvement projects. What will you do if the work doesn't meet your expectations and needs to be redone? Additional expenses and hard feelings are often the result. Considering how detailed the carpentry work will be may help you to decide. Knowing the experience, workmanship and dependability of the person may also help you to determine whether or not to hire him.

Q.

Do all contractors have to be licensed?

A.

No. Many states don't require their contractors to be licensed. Some states require contractors to be registered. Check your state requirements and obtain copies of all required insurance documents. Local requirements may differ. Call your building department and ask.

Q.

What if I use a contractor who is uninsured?

A.

In the event you hire an uninsured contractor, you may become legally responsible for injuries to persons and/or damage to personal property. Be sure all workmen on your project are covered under their employer's worker's compensation and general liability policies. Obtain copies of these documents before your project commencement date.

Q.

The general contractor I hired recently changed subcontractors. I requested license and insurance information from the new subcontractor and he told me he isn't licensed or insured. What should I do?

A.

Speak to the contractor. Most contractors carry insurance that covers all subcontractors. Also, the particular subcontractor may not be required to be licensed.

Q.

What is a Bond?

A.

Bonds are issued by surety companies. Their function is to guarantee completion of projects done by contractors. There are different types of bonds available. Bonds are generally posted for amounts equal to or greater than the full amount of the contract sum for a home improvement project. Some states' bond requirements, however, are for smaller amounts. Payment and performance bonds usually offer the best coverage. Confirm the amount of coverage before signing the contract.

Q.

How can a Bond protect me?

A.

In the event a contractor does not complete your project, the bonding company may be required to reimburse you the remaining costs or pay a contractor to finish the project.

Q.

Can any contractor get a Bond?

A.

If he qualifies. To qualify, a contractor must submit an application and supply the bonding company with license, insurance and financial information that meets required standards.

Q.

Should I require that the contractor I choose be Bonded?

A.

It depends. There are many points to consider, including the size of your project and the amount of money involved. Many states require bonds. Certain municipalities require a contractor to post a bond if he is not in business locally. If bonds are required in your area, ask for copies of the documents. Check them carefully for contents and validity.

Q.

How long are warranties on workmanship?

A.

One year in most cases, although they may vary. Roofing contractors generally guarantee their work for five to ten years. You may be able to negotiate a longer warranty period.

Q.

How can I be assured my contractor will show up regularly and keep my project going?

A.

Consider putting a stipulation in your contract stating that the contractor is to personally supervise and coordinate all facets of the project continuously until completion. Specify a set number of days per week that he must be on site.

Q.

I'm considering the purchase of a tract home in a development community. The sales agent said all I need to do is select the model I want and sign the purchase agreement. Any suggestions?

A.

Remember specifications! If the purchase contract does not specify every detail to your satisfaction, don't sign. Model homes are loaded with expensive upgrades and options. Make sure you know exactly what you will be entitled to receive.

Q.

I ordered a set of blueprints through the mail. Can a contractor use them for my project?

A.

In most cases the answer will be yes, however, the contractor will need to have them approved and sealed by a state licensed architect or engineer. Also confirm that the blueprints will meet all local building code standards. Ask the mail order company or a local architect about alteration fees if changes become necessary.

Q.

How can I find out about the lien laws applicable in my state?

A.

Contact your local building department — most provide brochures on lien laws. Also visit your local courthouse library. Check your state statutes, construction laws, real estate liens and mechanics' liens.

Q.

How long does a company who supplies materials for my remodeling project have to send me a Notice to Owner?

A.

Usually 45-60 days. The amount of time may vary from state to state, and should be calculated from the time the materials were received, not the order date. Some states don't require suppliers to send Notices to Owners.

Q.

Should a contract between a homeowner and builder be witnessed?

A.

Yes. Many contracts are required to be notarized.

Q.

Is it okay to use a commercial builder for a residential building project?

A.

Check with local authorities to make sure it is allowable in your area. If it is allowable, and the builder is reputable, you may want to use him. Ask the builder to show you (and let you inspect) recently completed projects he has done that are similar to yours.

Q.

The builder I would like to hire has provided me with a written estimate. He says we can sign the actual contract on the day he starts my kitchen remodeling. Is this a common practice?

It shouldn't be. You should request at least one week to review any type of contract, especially for a substantial amount of money. Also, if you want an attorney to review the contract and consult with you, chances are you won't be able to do it the same day.

Q.

A bathtub was installed in my bathroom that is not the type, brand or model that I specified in my contract. It was installed while I was at work. What should I do?

A.

Notify the contractor of the mistake. If the contractor takes responsibility and agrees to replace it *at no charge*, you are not likely to have a problem. However, if the contractor refuses to replace it, consider sending him a certified letter. State that you want the tub changed within a set number of days. If at this point he still refuses to change the tub, you might want to consider consulting with an attorney.

Q.

I paid a lien that was placed against my home. The lien was not valid to begin with. Can I get my money back?

A.

This is another question best addressed by legal counsel. If there is evidence that a lien has been placed against your property mistakenly, consider approaching the lienor before making any payment. If you have already made payment, request a refund from the lienor.

Q.

I learned about completion bonds after signing a contract with a builder. Can I require the builder to provide one now?

A.

It never hurts to ask, however, the builder may not be legally required to post a bond for your project. If he wants you to be a satisfied customer, however, he may consider it. Or you may be able to negotiate sharing the cost of the completion bond with the builder.

Q.

Is it okay to hire a contractor who specializes in bathrooms to remodel a kitchen?

A.

Before hiring any contractor who specializes in a certain area of construction it is a good idea to ask to visit previous projects. Try to visit jobs that are similar to yours. Ask questions of the owners regarding the quality of the project (while keeping in mind that their idea of quality may not be the same as yours). After researching the contractor's credentials and seeing his work, you may be able to make the best decision.

Q.

In the state where I live, a contractor must be licensed only if a building project contract exceeds $25,000. What happens if the original contract starts out below the set dollar amount, but later exceeds it because of change orders or extras to contract?

In most states, the contractor would be required to notify the licensing facility and tax bureau or regulatory agency of any change in the contract amount. Check your local licensing facility for this information (*see Chapter Four*).

Q.

Does every state require a permit to build or remodel?

A.

No. Most states do not require permits. Permits are regulated, and almost always required, by your town, city or county building department.

Q.

Can an architect acquire my building permit for me?

A.

Some states and municipalities allow architects to pull permits. Others require that an architect be licensed as a builder or contractor to pull permits. In some cases, an architect can act as the owner's agent and obtain a permit (this may or may not need to be in writing). *When an architect pulls a permit as the owner's agent, the owner may become the permit holder and be responsible for the project.* Usually only the contractor or builder should pull permits.

Q.

The addition I would like to have put on my home is very large. Will the rest of the home need to be brought up to code?

A.

Since local building and zoning departments regulate code requirements, you would need to check with yours. In many instances the homes are required to be brought up to code if the addition exceeds a certain amount of square footage.

Q.

The contractor I'd like to hire says my project will only take two weeks to complete. He's requesting a 50% deposit when I sign the contract and 50% upon completion of the job. I don't feel comfortable giving 50% up front. Should I?

A.

If the contractor needs to qualify for a building permit for your project, it may take as long as six weeks or more to obtain. Since headlines abound with stories of scam artists who have taken off with deposit money, consumers have become wary of even the most reputable contractors. Consumers may rightfully wonder where their money will be tomorrow. A suggestion might be to give a 10-20% deposit when the permit is in your hands and the work physically begins. Then regulate how much you pay as your job progresses. Try to avoid paying for more than the job is worth to date. Keep in mind that some states regulate draw payments.

NOTES:

Index

H

I

Z

About the Author

Steve Gonzalez is a Florida State Certified Residential Contractor with over 20 years experience in the building industry. During his career he has built over 100 homes and successfully completed over 100 home remodeling projects.

Currently designing and building custom single family homes in South Florida, Steve writes many articles, reports and columns on working with contractors and project planning. He has been featured nationally on television and radio, and is frequently a guest speaker at home shows across the nation.

Order Form

Copies of ***Before You Hire A Contractor*** are only $12.95, plus $2.50 shipping and handling. Florida residents add 6% sales tax.

Name____

Address______________________________________

City________________State______________Zip______

___ copies @ $12.95 =	_________
6% Sales Tax (FL res. only)	_________
Shipping/Handling ($2.50 first copy, $2.00 each add'l copy)	_________
TOTAL	_________

Mail your order to:

CONSUMER PRESS, INC.
Order Department BHC
13326 Southwest 28th Street
Ft. Lauderdale, FL 33330-1102

☐ Please add my name to your mailing list for upcoming titles.